nce Henry the Navigator

By the same author

CUCUMBER SANDWICHES IN THE ANDES

Statue of Prince Henry at Lagos in the Algarve

JOHN URE

Prince Henry the Navigator

CONSTABLE LONDON

First published in Great Britain 1977 by
Constable and Company Ltd
10 Orange Street London WC2H 7EG

ISBN 0 09 461240 4

Set in 12 point Monotype Bembo
Printed in Great Britain by
The Anchor Press Ltd and bound
by Wm Brendon & Son Ltd
both of Tiptree, Essex

For my god-daughter

CATHARINE

who is already an experienced traveller
in Portugal and Africa

Contents

Illustrations

MAPS

Prince Henry's coat-of-arms on the title-page and the diagram on page 101 were drawn by Miss Rosemary Greenfield

Preface

For the past three years I have lived in Lisbon in a house with one of the most evocative views in Europe. From the terrace one looks across gardens, tree-tops and old roofs onto the River Tagus at the point where the stream swells out into a mighty inland pool, in which shipping from all over the world lies at anchor: ocean liners, jumbo tankers, warships and tiny, dhow-rigged river craft are all to be seen. A few miles downstream lies the old port of Belem, with its yacht harbours, its great monastery church of St Jeronimo and its massive monument to Prince Henry the Navigator and his companions. No one could live here for long and be forgetful of Portugal's role in opening up the seaways of the world.

As I have travelled around Portugal, this awareness of her past has been illumined and enhanced by her present. From the steep, narrow lanes around the riverside at Oporto one stumbles on the birthplace of Prince Henry himself; the castle of Tomar still preserves its links with the Order of Christ under whose banner the early Portuguese navigators sailed; the fishing village of Sines (now in the process of transformation) still displays the statue of its most illustrious son – Vasco da Gama; the ancient Palace of Cintra still preserves the rooms where Prince Henry, with his father and brothers, planned Portugal's first overseas venture; and – above all – at Sagres, in the Algarve, the rocks remain as a natural monument to the man who lived there and devoted his remarkable energies to initiating Portugal's great era of discovery.

Travelling was supplemented by reading. The more intrigued I became by the fifteenth-century discoverers, the more my interests focused around the enigmatic figure of Prince Henry himself. The British have always felt a special affinity for this half-English Prince. It was no surprise to find that many of the most authoritative studies of him were by Englishmen: Richard Henry Major had written a massive volume over a hundred years ago, and had been followed by Beazley, Prestage, Bradford and others. To all of these – in varying degrees – Prince Henry was a hero, a person ahead of his time who was both a scientist and a 'modern' man.

But when I turned to recent or contemporary Portuguese historians I found some very different verdicts. Oliveira Martins portrayed a man who compared badly with his brother, Prince Pedro; who was, in fact, a blinkered and less than noble figure. Professor Magalhães Godinho – that notable contemporary Portuguese historian – was even more disturbing: to him the great achievements of the Portuguese discoveries were not the legacy of Prince Henry nor indeed of any individual or group; they were the results of the inexorable working of economic forces.

I turned my attentions to the early chroniclers and read their own accounts; I visited the national archives at the Torre de Tombo, the libraries of the Lisbon Geographical Society and the Academy of Sciences, and other records elsewhere in Portugal. Eventually I found I could subscribe to neither of the accepted views. Prince Henry appeared to me a much more complex figure than the precursor of the Renaissance portrayed by Major; he seemed a figure moulded in the medieval world and possessing many of the rigid mental attitudes of that world. And yet his personal rôle in initiating the era of discovery seemed to me indisputable and paramount; for me it could not be reduced to a facet of larger political or economic forces.

Gradually I began to evolve my own view of a man torn between the conflicting influences of a medieval background and upbringing on the one hand, and a pragmatic and forward-looking personality on the other. The various contradictions in Prince Henry's behaviour and the inconsistencies in his career appeared more explicable once this fundamental dichotomy was recognized. I set out to retell his story in this light.

At every hand I encountered encouragement. The staffs of the Lisbon libraries mentioned above were patient with me. Professor Magalhães Godinho received me with courtesy and erudition. Captain Teixeira da Mota allowed me to question him on those aspects of fifteenth-century seamanship on which he is probably the greatest living authority. Those three wise men of Anglo-Portuguese scholarship – Professors P. E. Russell, C. R. Boxer, and George West – were kind and constructive in their comments when they heard of my endeavours. Dr Carlos Estorninho (of the British Institute in Lisbon) was unflagging in his efforts to locate and obtain books for me. My *cher collègue* in the Austrian Embassy – Dr Jörg Schubert – went to great trouble to take some of the photographs that illustrate the book.

Yet even with so much encouragement and help I became increasingly aware of the temerity of my undertaking. As a professional diplomat living through some of the most traumatic and exciting few years of Portuguese history to have occurred during the five centuries since that period about which I was writing, I had little time and often less energy remaining for my self-imposed task. I was aware of the weight of scholarship that had preceded me and seemed likely to succeed me in my enquiries. But my absorption in Prince Henry increased as my researches continued; there was refreshment to be found in the study of the life of one of the greatest sons of that nation among whom I was living and for whom my affection ever grew. If this is a personal view of Prince Henry, it is at least one which – I hope – may kindle in others the fascination which I have felt for a complex character whose forcefulness and dedication contributed so greatly to one of mankind's most notable achievements.

Note on Portuguese names

I have tried to be consistent about the spelling of any particular individual or place throughout this book, but I have not achieved any overall consistency. For instance, it has seemed to me pedantic to refer to Prince Henry as Henrique; and yet to have referred to his brothers Duarte, Pedro and João, as Edward, Peter and John would have appeared to me to be excessive anglicizing. Similarly, I have preferred Leonora to Leonor and Braganza to Bragança, but have left most other names (within Portugal) in their Portuguese forms. Even where a Portuguese spelling has been adopted, this has not been the end of possible variations. Many such names have older forms with double letters, such as Ean(n)es, Barcel(l)os, Af(f)onso; here I have adopted what seemed to me the most customary form in each case. I have preferred Cintra to Sintra and Zurara to Azurara, for reasons which may appear arbitrary. Where my decisions are open to correction I can only plead that I have tried to act in the interests of clarity.

Chapter 1

Portugal in the World of 1394

'This nurse, this teeming womb of royal kings,
Fear'd by their breed, and famous by their birth,
Renowned for their deeds as far from home, –
For Christian service and true chivalry . . .'

Shakespeare: John of Gaunt in *Richard II*

All times are times of transition to those who live through them. But the year 1394 – in which Prince Henry of Portugal was born – had more claim than most years to be considered as a period when the world was on the move, when new forces were replacing old ones, when new values were challenging accepted ones, when new ideas were penetrating established patterns of thought.

Not everywhere were the new forces progressive ones. In China the descendants of Genghis Khan and the Mongol emperors, who had been the hosts to Marco Polo exactly a hundred years before, had recently been defeated by the first of the Ming dynasty emperors, and an era of isolation from Europe had begun. On the steppes of Russia, the Golden Horde of the Tartars was suffering its final defeat at the hands of the equally destructive Tamburlaine. In Asia Minor, the Ottoman Turks were advancing across the Anatolian plains towards the rich prize of Constantinople, which – with its Byzantine treasures and irreplaceable records – was to fall to the Saracens' attack half a century later. In north-west Europe, the Hundred Years' War between France and England was experiencing one of its uneasy truces but was soon to be resumed with fresh vehemence.

But elsewhere the new forces were more creative. Above all on the Italian peninsula something remarkable was astir: the Renaissance was already turning men's minds into fresh channels.

It was nearly a century since Giotto had completed his frescoes of the life of St Francis at Assisi, and European painting had been given – almost literally – a new dimension. Already in Florence the infant Fra Angelico was learning to hold a brush and the young Donatello a chisel. The new movement in the graphic arts was to spread into literature, politics and philosophy, and was to replace alchemy with science, superstition with experiment, prejudice with logic. By 1394 Petrarch and Boccaccio were no longer alive, but the literary ripples that went out from the Italian city states were reaching distant parts – were reaching, for instance, to London and Kent where Geoffrey Chaucer (enjoying the fitful patronage of King Richard II) was already in the throes of writing his *Canterbury Tales.*

And in that other great European peninsula – Iberia – life was also on the move. The Moorish occupation of the peninsula was nearing, but had not yet reached, its end. At Granada the Sultan Mohammed V still presided in elegant security at his Alhambra palace, where he had just completed his celebrated Court of the Lions. Castile and Aragon were still ununited, while to the westward the newly consolidated Kingdom of Portugal, which had thrown off the Moorish conquest more than a century before, was rousing itself for its golden age.

But it was a very tiny and feeble kingdom. The Black Death of 1348–9 had reduced the population of Portugal to well under one million people, who were spread thinly throughout the countryside rather than concentrated in cities or towns. Lisbon with 40,000 inhabitants was far ahead of Oporto with 8,000, which in turn had more than twice the population of other centres such as Coimbra, Evora or Braganza. Within these larger towns life was pursued on a strictly medieval pattern: trade guilds controlled the work of the artisans who tended to congregate in quarters reserved for their own particular 'mysteries'; Jews or other aliens would be herded into ghettos; the bells of the numerous private chapels and churches would continually remind the citizenry of the wealth, influence and ubiquity of the Catholic Church.

In the countryside, smaller towns and villages were perched on hilltops behind their crenellated walls, separated from each other by vast tracts of forest or scrub – frequently infested with rogues and robbers. The roads were everywhere appalling and travel from one town to another was both dangerous and arduous. In the cultivated areas, the peasants grew wheat, grapes and olives depending on the nature of the soil. In the south they tended to labour on large estates, while in the north many of the peasants rented their own smallholdings. All were liable to serve in the King's army if the realm were invaded, and even in peacetime were required to pay heavy dues to the Crown, their feudal lords and the Church – such payments not infrequently amounting to as much as seventy per cent of all they produced. Many dues and commercial debts were settled in kind, partly because the currency was inadequate for the nation's needs: no gold coins were minted in Portugal between 1385 and 1435, and the silver coinage was continually debased by admixtures of copper.

In the coastal regions of Portugal a lively fishing industry flourished, despite a lack of natural harbours on a coastline which was – for the most part – open to the rigours of the Atlantic. A modest mercantile marine business was also conducted, particularly with the ports of northern Europe where Portuguese wines, hides, fruit and olives were exchanged for grain, cloth, iron and gold coin. The Portuguese had already sensed that the sea was important to their national life. This, then, was the nation which was poised on the brink of its golden age. The immediate prelude to that golden age was brutal, bloody and vicious, but some account of it is necessary to an understanding of the life of the young Prince who was born on Ash Wednesday the 4th of March 1394.

Prince Henry's grandfather had been King Pedro I of Portugal, who was known as Pedro the Severe. His lifetime and reign (1357–67) were beset by intrigues and disputes between the powerful Castile and the precariously independent Portugal. It was a turn in this dispute that was responsible for the first of a violent chain of events which was to set the style of barbarism in the years preceding Prince Henry's lifetime. If the account in the immediatley following pages seems to dwell excessively on this lurid side of events, it may serve the better to contrast with the relative calm and civility in which – only a decade later – the young Prince Henry was to grow to manhood.

Pedro had fallen in love with a lady-in-waiting of his wife, by the name of Inês de Castro. After his wife had died in childbirth, Pedro either married or attempted to marry (the controversy as to which involved correspondence with the Pope and has never been completely resolved) Inês de Castro. Pedro had not yet succeeded to the throne of Portugal and his father – King Afonso IV – thought that the family of Inês de Castro were intriguing with Castile. He therefore authorized her murder* at a place subsequently called the Fountain of Tears, near Coimbra. Pedro consequently took up arms against his father, but only temporarily and ineffectually; when he succeeded to the throne shortly thereafter he seized the murderers, had their hearts torn from their living bodies, disinterred the body of Inês de Castro and made the whole court kiss her lifeless hand in token of her sovereignty.

This macabre ritual was fresh in memory when Pedro was succeeded by the son of his first wife – Fernando I – who in due course also inherited a claim to the throne of Castile. The claim was disputed and Fernando sought the support of Aragon by engaging to marry the King's daughter, Leonora; simultaneously the Pope intervened with a solution involving his engagement to the Castilian heiress – also named Leonora; meanwhile, Fernando himself had developed a passion for the wife of a nobleman at his court – she too was called Leonora (and surnamed Teles). King Fernando chose the worst possible course. He set aside the Aragon engagement, thus jeopardizing an alliance. He rejected the Castilian Leonora, thus provoking war with Castile. He annulled the marriage of Leonora Teles (sending her husband abroad), thus saddling himself with a wife of ruthless ambition

*Recent scholarship suggests that she may have been granted some form of hasty trial for high treason and that her death could be described as an execution.

Simplified Genealogy of members of the Portuguese Royal Houses

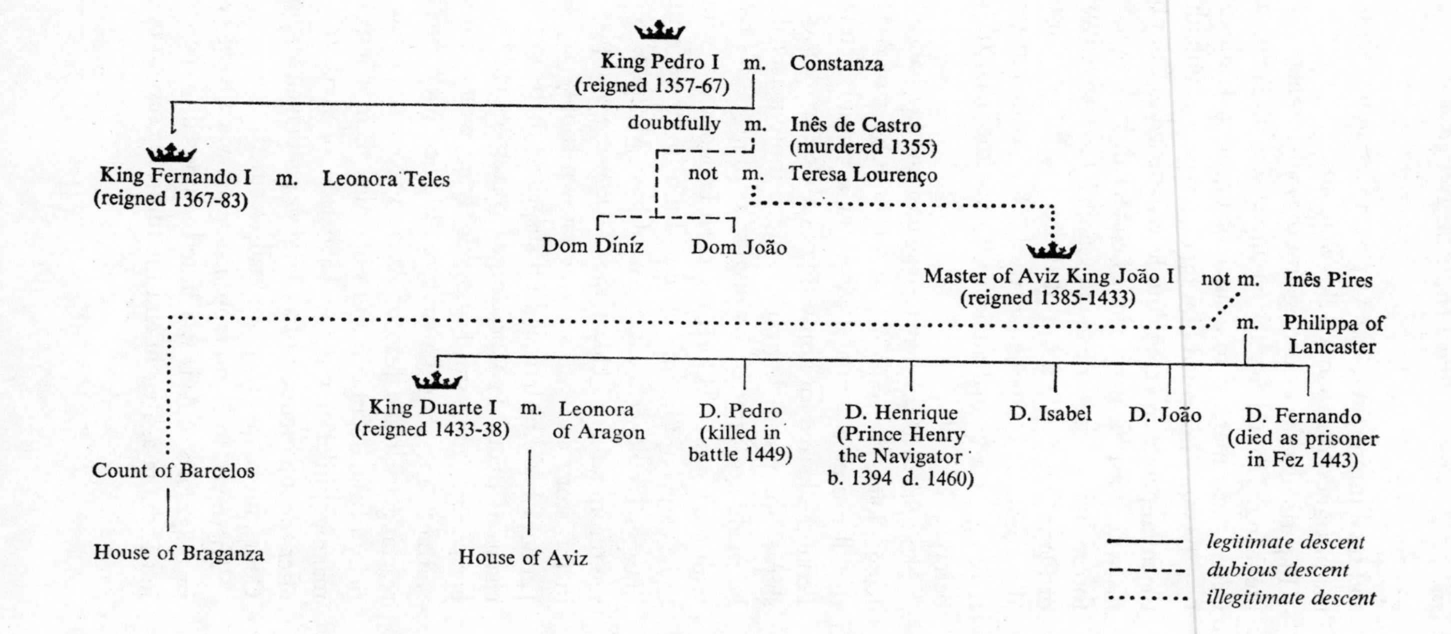

and total immorality. The war with Castile was soon ended, but the intrigues of Leonora Teles were not.

In particular she feuded against the half-brothers of her husband the King. Dom João, the younger son of Pedro and Inês de Castro, was married to a sister of Leonora. Undeterred by this fact, Leonora informed him that his wife was unfaithful to him; the intemperate and credulous João hastened to Coimbra and slew his wife with his own hand. Leonora then denounced him and forced him to flee to exile in Castile. Thereafter it was the turn of Dom Diniz, also a son of Pedro and Inês de Castro and thus also considered a potential rival. He was driven into exile for refusing, at a formal audience, to kiss the hand of his brother's betrayer, Leonora.

Next she turned her malice against another half-brother of the King – João, the Grand Master of the Order of Aviz – who was the illegitimate son of Pedro and a certain Teresa Lourenço. Leonora's hatred for him was not only on account of his relationship to the throne but on the much more personal ground that he made no secret of his disgust at the Queen's open adultery with the Count of Ourém. Indeed Leonora was widely believed to be carrying Ourém's child and thus to be deflecting the royal lineage. To silence the criticism of João, Master of Aviz (who – as the future father of Henry the Navigator – is to play a large part in this story and will now be known simply as João of Aviz), Leonora had him arrested on suspicion of treason and held in chains at Evora. Leonora forged correspondence to substantiate her charges and – still fearing the King would not sign a death warrant – she forged that too. But her wickedness was overreaching itself and João of Aviz's gaolers suspected the authenticity of the document and suspended their action. Leonora was making futile attempts to have João of Aviz poisoned when, after the intervention of an English gentleman at the Portuguese Court, his release was eventually secured.

There was to be no rapprochement when King Fernando died in 1383. João of Aviz was bitterly opposed to the settlement that left Leonora as Regent and the future succession bound up with her descendants and the royal house of Castile. A tense interview at the palace at Lisbon between Leonora and João of Aviz, at which she offered him the governorship of the Alentejo to remove him from the capital, ended by João of Aviz leading Leonora's lover the Count of Ourém into an anteroom and murdering him within earshot of the Queen Regent. (The incident has parallels with Darnley's murder of Riccio within earshot of Mary Queen of Scots.)

The killing had been premeditated and João of Aviz had laid his plans well. He gave orders that the palace gates should be closed and his page rode through the city proclaiming that his master, João of Aviz, was once again Leonora's prisoner. A popular uprising followed, during which a crowd of João of Aviz's supporters stormed the palace and Leonora only escaped with difficulty, to raise her standard first at Alenquer and then in the greater safety of Santarem Castle. While the Portuguese aristocracy rallied to her, the merchant classes in the cities rallied to João of Aviz. Leonora appealed to the King of Castile, who marched an army into Portugal in her support. João of Aviz appealed for help to England, although – so bereft was he of aristocratic supporters – he had difficulty in finding anyone of sufficiently noble birth to plead his case in an acceptable way at the rank-conscious English Court. Foreign invasion and civil war were once more the fate of Portugal.

The following two years were spent in preparation for a final test of power. João of Aviz regularized his position by undermining the claims of his two half-brothers, Dom João and Dom Diniz, to the Portuguese throne. He maintained that the marriage of their father (who was also João of Aviz's father) to Inês de Castro had not been valid; so, in fact, they were as much bastards as he was. The process of discrediting them was made the easier by the fact that the two Princes had been imprisoned by the King of Castile – where they had both taken refuge after their persecution by Queen Leonora – and so were unable to plead their own case. Even so, some heavy-handed support by his friends – including that of a formidable soldier named Nun' Álvares

Pereira who was to feature largely in subsequent events – was necessary before João of Aviz was proclaimed King João I of Portugal by the Cortes in Coimbra in March 1385. Now his task was to defeat the Castilian invader and establish himself as King in reality as well as in name.

King João's envoys to England, despite their plebeian origins, had met with considerable success and two English ships carrying a contingent of the famous English long-bowmen arrived at Lisbon in April. John of Gaunt, Duke of Lancaster, had long been pressing his claim to the Castilian throne and saw in João of Portugal a useful ally. The King of Castile, for his part, had received reinforcements from France.

The decisive clash came on the 14th of August 1385 at Aljubarrota. King João's army probably numbered some 3,000 men-at-arms and about as many again of local militia, reinforced by some 500 English bowmen. The Castilian army was almost certainly considerably larger, possibly twice the number. Nun' Álvares was in command of the Portuguese vanguard and chose his position with skill and care. The weary and ill-disciplined Castilian army were forced to make a flanking detour in order to improve their position. While they were effecting this manoeuvre the Portuguese constructed brushwood obstacles and dug ditches so that even the new Castilian position was an unattractive one from which to launch an attack. In these circumstances, the King of Castile and his council wisely decided to postpone action and allow hunger to force the Portuguese army from their strong position. But the royal orders were disobeyed by the headstrong Castilian *hidalgos* who launched a premature charge which led on to a full-scale engagement. The English archers took a heavy toll of the Castilian cavalry but, even so, the Portuguese front line reeled under the weight of the survivors. King João threw in his main force to support the front line. The fighting was fierce but relatively short. Within an hour the Castilian royal standard had fallen and their King was given a fresh horse and advised to flee the field. The rest was a rout. It was not only the Castilians and their entourage (including an unfortunate French Ambassador) who were killed, but a great number of the Portuguese aristocracy who had been fighting on the Castilian side and against the upstart João of Aviz.

Modern historians consider Aljubarrota one of the decisive battles of medieval Europe. If it did not ensure the permanent independence of Portugal, at least it can be said that a contrary result of the battle probably would have ensured a permanent end to that independence. King João emerged as undisputed master of a united kingdom – a monarch whose hand the rulers of Europe might covet for their daughters. But there was a small problem regarding a royal marriage: as Grand Master of the Order of the Knights of Aviz, João had been sworn to chastity. However, he had never taken this injunction too literally, as the early installation of Inês Pires as his mistress bore witness, and he quickly sent an emissary to Rome to ask for a dispensation from the Pope to enable him to effect a suitable alliance.

The approach which prompted these efforts to establish his eligibility was from none other than John of Gaunt, Duke of Lancaster, his English friend and ally. Lancaster's claim to the Castilian throne gave him an interest in cementing an alliance with the Portuguese King and, following the victory of Aljubarrota, he offered João a choice of either of his daughters. João wisely chose the elder – Philippa – who was devoid of those dynastic claims which her younger half-sister had inherited from her Spanish mother.

The marriage of Henry the Navigator's parents was celebrated in Oporto on St Valentine's Day 1387, without the Papal dispensation. The fact was that Lancaster was in such a hurry to launch a military campaign to realize his claim to Castile that he rushed his daughter and King João to the altar without even giving the latter sufficient time to collect the notables of his realm around him. The Archbishop of Braga scarcely had time to preside over the torch-light bedding ceremony in the bridal chamber before messages about the marching orders for the invasion of Castile started arriving from the Duke of Lancaster. The bride's father had curiously absented himself from the wedding ceremony, probably for the cynical reason that – while he wanted Portuguese soldiers – he did not want to offend the susceptibilities of his future Castilian subjects by unnecessary junketing with the Portuguese Court.

Less than a week after the wedding, King João was on the road with his army, although it was mid-March before the English and Portuguese forces united at Braganza. The English army had been weakened by pestilence on the march from their landing point at Corunna and were reduced to about 1,500 men, as opposed to the Portuguese contingent of 9,000. Thus it was hardly surprising that Nun' Álvares – now Constable of Portugal and the chief of King João's generals – insisted on the command of the vanguard. The campaign which followed was to test Portuguese forbearance with their allies rather than courage against their adversaries.

It might have been expected that Lancaster, who had been impatient to begin, on whose behalf the campaign was being fought, and who was so beholden to his ally's goodwill for the main part of the invasion force, would have wanted to press on with vigour and single-mindedness – if only so that his Portuguese allies should not weary of the campaign. He did nothing of the sort. As each successive Castilian town was reached and besieged, Lancaster sent out heralds to enquire the names of the principal knights involved on the enemy side. If these included any gentlemen of renown – particularly any French knights with whom he had done battle before – Lancaster would call for a truce and organize individual jousting combats in front of spectators from both armies.

This display of chivalric *camaraderie* both exasperated and fascinated King João and his Constable. They were used to the rough and ruthless ways of the Iberian peninsula where war was a serious affair, and they were acutely conscious that at any moment Castile might receive decisive reinforcements from France. Time therefore was not to be frittered away. On the other hand they were entranced by the nonchalance and panache of these English and French amateurs who made an art of war. Nun'

Álvares had long been conversant with the tales of King Arthur's Knights of the Round Table: from now on he was thought by many of his friends to have identified himself with Sir Galahad! King João was to be much influenced in his later years by memories of his patrician father-in-law's style.

The campaign was a total fiasco. The reluctance of the English commanders to take it seriously (at Salamanca they dined with their French opponents off wine and meat supplied by the latter while their Portuguese allies went hungry) and the lack of siege equipment on the Anglo-Portuguese side necessitated a retreat and the final abandonment of the campaign. Lancaster bade farewell to his daughter, Philippa, and embarked for England, having arranged a marriage for another of his daughters to one of the very Castilian princes whom he had so recently been fighting. It had been a curiously inauspicious start to six centuries of Anglo-Portuguese alliance.

King João was left bemused. The situation of Portugal was neither weaker nor stronger than before the campaign: he retired behind his frontiers to consolidate his position. And one method of consolidation was to beget heirs. In 1388 his Queen, Philippa, gave birth to a daughter who died after eight months. In 1390, she gave birth to a son who died after two years. In 1391, she gave birth to Duarte – later to succeed his father on the throne. In 1392, she gave birth to Prince Pedro. And then in 1394 (with three children to follow) she gave birth to Henrique – to be known to Englishmen as Prince Henry the Navigator.

Chapter 2

A Chivalric Court

'Now thrive the armourers, and honour's thought
Reigns solely in the breast of every man:
They sell the pasture now to buy the horse;
Following the mirror of all Christian Kings . . .'

Shakespeare: Prologue to Act II of *King Henry V*

The Court in which Prince Henry grew up was not by tradition a chivalric one. Quite the contrary: the Portuguese monarchy had been nurtured on the rough manners and the haphazard morality of a military camp. But during the years of Prince Henry's adolescence the court was self-consciously determined to become chivalric – to acquire that whole paraphernalia of complex mental attitudes and modes of behaviour which had inspired aristocratic society in the more sophisticated countries of fourteenth-century Europe. The young Prince immediately responded to these courtly influences and they were to become one of the twin strands in his character throughout his life.

Throughout Europe, the whole concept of chivalry had become a curiously elaborate one. No longer was it – even if it had ever been – a simple convention by which the higher social orders refined the more brutal aspects of the Middle Ages. That convention – largely based on Arthurian legend – had had certain clearly distinguishable features: the knight had an obligation to protect – and not merely exploit – the weak; casual lechery was replaced by respect for – and sentiment regarding – women (or at least ladies); dishonourable conduct – such as the violation of a truce or safe-conduct – discredited the perpetrator. But out of these relatively straightforward concepts grew an elaboration of ever more rarefied social behaviour and convoluted moral reasoning. There were well-documented instances of knights wearing golden shackles, covering one eye in battle, or refusing to sit down to eat or drink, until some objective of 'honour' had been achieved. Protocol became all important: knights bearing urgent messages in battle to their commanders in the Hundred Years' War would delay delivering them until they had established the precedence of the messengers.

The eminent Dutch historian Huizinga* described all aristocratic life in the later Middle Ages as an attempt 'to act the vision of a dream'. Indeed by the early fifteenth century there was an almost total unreality about the set of values by which the chivalric Courts of Europe strove to regulate their activities, and the greatest unreality of all was the self-deception which they practised regarding their own motivation. Huizinga observed that the chroniclers of the chivalric Courts told a tale of treason, cruelty, calculated self-interest and subtle diplomacy; yet they rationalized these events as motivated by high knightly concepts. The grisly histories of Froissart are a case in point, as are the writings of the Burgundian chronicler Chastellain, who genuinely believed that the magnificence and opulence of the Burgundian house was based on the dashing exploits of its knight errants rather than on the solid commercial prosperity of the Flemish and Brabant merchant towns.

Had Huizinga turned his attention further west – to the Portuguese Court – he could have pointed out the same incongruities in the fifteenth-century writings of Zurara, the main chronicler. Zurara saw everything he recounted in a chivalric context, whether it was a trial of strength with the Moors at Ceuta or a squalid nocturnal scuffle between Portuguese seamen and negro tribesmen on an African beach. For him, war was the criterion of nobility; when writing about the inhabitants of Teneriffe, for example, he commented that 'they live more like human beings than those of the other islands . . . their chief care is to fight one another'. His belligerency reaches even greater absurdity when describing the inhabitants of Grand Canary, whose savage chiefs

*In *The Waning of the Middle Ages.*

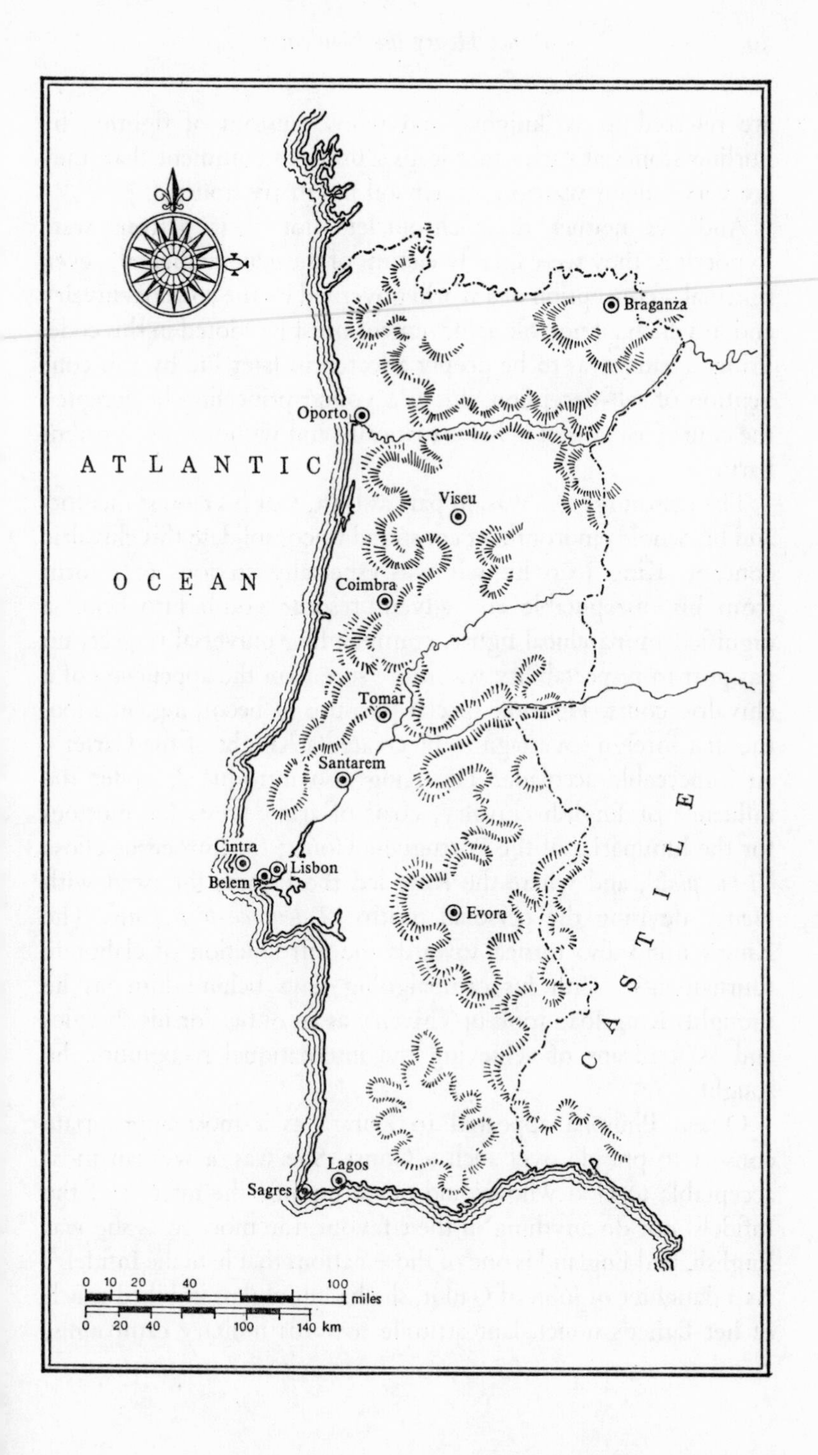
Braganza
Oporto
ATLANTIC
Viseu
OCEAN
Coimbra
Tomar
Santarem
Cintra
Lisbon
Belem
Evora
CASTILE
Lagos
Sagres
0 10 20 40 100 miles
0 20 40 100 140 km

are referred to as 'knights' and whose custom of fighting by hurling stones at each other leads Zurara to comment that 'they are very valiant warriors – their soil has many stones'.

And yet neither these chroniclers nor their patrons were hypocrites: they were merely victims of the general mental – even spiritual – conception of a world governed by the code of chivalry and in which all notable achievement must be rooted in this code. Prince Henry was to be deeply affected in later life by this convention of self-deception, but as a young princeling he accepted the courtly canons of belief uncritically and without any apparent harm.

The reason for this was, in part at least, that his closest mentors and his whole environment conspired to consolidate this chivalric concept. King João himself was especially anxious to reform from his disreputable and adventuresome youth into being a dignified monarchical figure, commanding universal respect; no passport to respectability was more sure than the appendage of a chivalric court. He took special pleasure in becoming, in 1400, the first foreign sovereign to be created a Knight of the Garter – an impeccable accolade. The King also introduced, under the influence of English chivalry, coats-of-arms, crests and mottoes for the luminaries of the Portuguese Court; for himself he chose '*Il me plaît*', and where the King led the Princes followed with Henry devising the personal motto '*Talent de bien faire*'. The King's mind also turned towards the introduction of elaborate tournaments. With his campaigning days behind him (as he thought) King João took up chivalry as an outlet for his energies and as a means of achieving the international recognition he sought.

Queen Philippa appeared to Zurara as a most appropriate consort to preside over such a Court. She was 'a woman most acceptable to God who would never espouse the interest of the Infidels, nor do anything in their favour, the more so as she was English, and England is one of those nations that hate the Infidels'. As a daughter of John of Gaunt, she had doubtless imbibed much of her father's nonchalant attitude towards military campaigns, though she had imbibed none of that disregard of sexual morality which had led him to install his mistress in the same palace as his wife, and pay conspicuously greater attention to the former. Indeed, for Queen Philippa chivalry was a very personal extension of religious discipline; she utterly rejected those aspects of courtly life which drew their inspiration from the varying secular themes of the *Roman de la Rose* or which glorified the relationship of the knight to his mistress. With her years of rapid child-bearing behind her, Queen Philippa set about moulding the court in her own image. The nineteenth-century Portuguese historian, Oliveira Martins, wrote that 'she found the Court a sink of immorality; she left it as chaste as a nunnery'. The achievement was not effected without the Queen having to exert all her influence over her husband, in whose own concept of chivalry morality played a much more subordinate part. His own mistress had early been despatched to a nunnery and any lapses into former habits on his part were sharply chided by the Queen.* More drastic fates were to await other deviants from the Queen's concept of knightly behaviour: Fernando Afonso, a young squire of the Court, who was detected in an affair with one of the Queen's maids-of-honour, and who declined to mend his ways after a warning, was arrested, escaped, sought sanctuary, was dragged from sanctuary on the King's orders, and was promptly burnt at the stake lest his conduct should contaminate the newly purged Court.

Not all the Queen's energies were devoted to domestic piety. It was she who commissioned the forging of swords for her sons and who presented them on her deathbed with solemn charges to observe the highest standards of the knightly code, Prince Henry himself being charged with the special care of the nobles, knights, gentlemen and squires of the realm. But undoubtedly to Henry, the revered figure of his mother must have coloured his vision of chivalry with unusually strong overtones of morality,

*There is a room in the palace at Cintra which is still commemorated as the scene of the Queen surprising King João in an illicit advance to a lady of the Court.

even perhaps – in the light of his own sexual abnegation – of chastity.

His elder brothers, too, contributed to the atmosphere of medieval chivalry in which Prince Henry came to manhood. The oldest – Prince Duarte – wrote a copious treatise on courtly behaviour (and, incidentally, on the practices and events of life at the Portuguese Court) entitled *O Leal Conselheiro* which, when contrasted with the Italian treatise on the same subject – Castiglione's *Il Cortegiano* – written nearly a century later, revealed how far the Portuguese Court still was from the spirit of the Renaissance. The second brother – Prince Pedro – was to spend much of his later youth in travelling round the courts of Europe, acquiring not only a Garter, like his father, but a thorough grounding in contemporary chivalric practice. Among those he encountered was King Henry V of England, who carried chivalric conceits to such lengths that, when on the way to the field of Agincourt he inadvertently rode beyond the village which his foragers had selected as the night's quarters, he refused to turn back because he was wearing his coat-armour and the traditions of chivalry forbade him to retreat when thus attired. (Subsequently, a special decree enjoined knights to shed their coat-armour when reconnoitring, to avoid this predicament.) Although Prince Pedro's travels lay in the future, his interest in foreign courts was pronounced even during his younger brother Henry's childhood.

Last, and perhaps strongest of all the personal influences on the young Prince Henry, was the shadowy but formidable figure of the Constable of the Realm – Nun' Álvares Pereira. This remarkable soldier, who had been beside the King in all the bitterest fighting that had led to the establishment and consolidation of the House of Aviz, was by now an acknowledged model not only of heroism but of rectitude. He – almost alone among the warring commanders of the Iberian peninsula – had disciplined his forces both to respect the possessions of the peasantry over whose villages they were campaigning, and to abstain from indiscriminately ravaging their wives and daughters. Although he lived away from the Court on his estates in the Alentejo, the

The Battle of Aljubarrota

The Royal Palace at Cintra in which the Ceuta expedition was planned

The Castle of Valença de Minho where some of Prince Henry's fleet was constructed

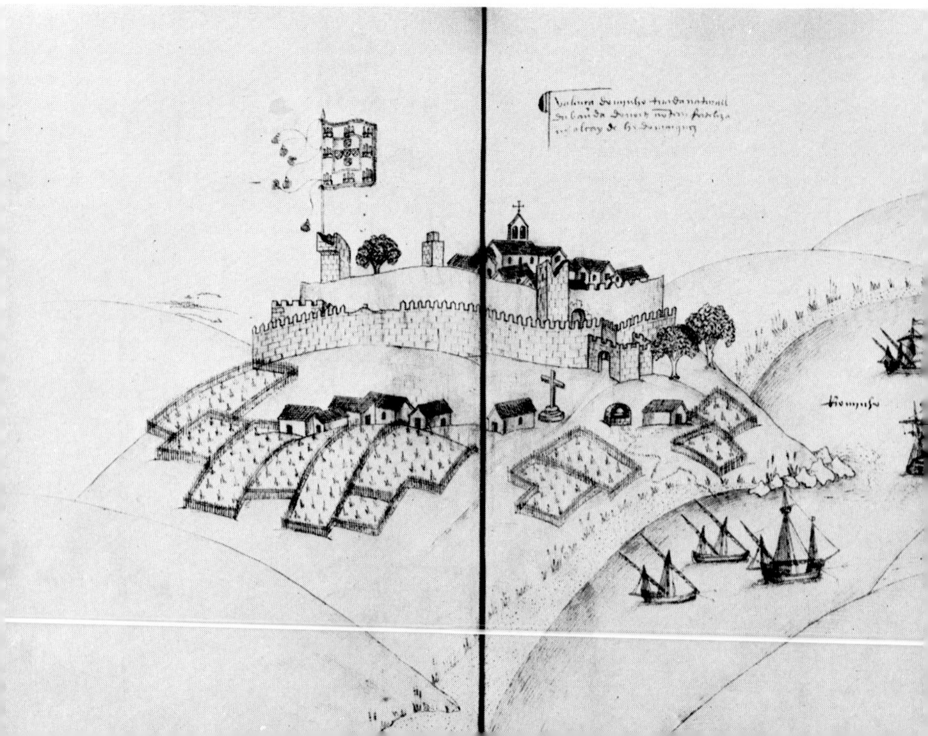

Constable's influence was widely felt and his reputation held up as a model to the Princes. On the first major venture of their lives, his support was to prove decisive.

But perhaps an even stronger influence on the young Prince Henry than any of these individuals was what might broadly be described as the crusading legacy of Portugal. The expulsion of the Moors from the Algarve was still fresh in legend if not in memory, and in neighbouring Castile the pennants of Islam still flew over Granada as a constant reproach to Christian knighthood. The four great military orders of St John, Santiago, Aviz and Christ still occupied their castles throughout the length and breadth of Portugal; Prince Henry's father had been Master of one of them, and he himself was to be deeply involved throughout his life with the most spectacular of all of them – the Military Order of Christ. This Order was a Portuguese derivative of the Knights Templar to whom had fallen the task in the preceding two centuries of keeping open the routes to the Holy Land and – in effect – of forming the storm troops of Christendom.

Prince Henry was consequently reared on the lore of the Templars. He would have learnt how they had been founded in 1118 by King Baldwin I of the Crusaders' Kingdom of Jerusalem to keep open the pilgrim routes to the Holy Land, and how they had expanded from being a motley band of poor – and sometimes excommunicated – knights to being an élite brotherhood, renowned for their gallantry, their rich endowments and their independence of all authority save only that of the Pope.

The young Prince must have been enthralled by the tales of how, in the two centuries that followed their foundation, the Templars' famous red cross on a white tunic was seen wherever Christian and Saracen forces met in fiercest combat. He would have heard of their arrogant and disastrous bravery at the siege of Ascalon where the collapse in flames of the Crusaders' wooden siege-tower – set on fire by the defenders of the city – caused the masonry of the walls to melt. Forty Templar knights had rushed through the gap, only pausing to place sentries on the *outside* of the breach to prevent reinforcements joining them and sharing in their glory; later in the day forty Templars' bodies were swinging from the parapets to rebuke such selfish bravery. He would have learnt how it was the Templars who urged the fatal march on Tiberius in 1187 which led to the destruction of the whole Crusader army on the Horns of Hattin. (Perhaps Prince Henry was to recall these incidents later when he was himself leading the assault on the Infidel stronghold of Ceuta or leading a crusading army to disaster across the deserts near Tangier.)

The pride of the Templars was as legendary as their courage: one Grand Master – Odo of Saint-Amand – when captured by the Saracens in 1180 preferred to die in captivity than be exchanged for a prisoner of less than princely status. (Did Prince Henry think of him much later when his own brother was lying captive and dying in Fez?)

Prince Henry would have read of the immense riches of the Templars, of how they became the bankers of Europe, lending ransom money for King Louis IX of France and lending dowry money for the daughter of King Philip IV; he would have read of their trade with the Moslem merchants of Egypt (did he think of it when he started bartering with pagans in other parts of Africa?) and of the terrible accusations of sodomy, blasphemy and witchcraft brought against them at the beginning of the century in which he himself was born. It was on Friday the 13th of October 1307 that the King of France had given orders for the arrest of all Templars throughout his domains, and had subsequently prevailed on the Pope to issue a Bull, outlawing the Order elsewhere in Christendom. He would have been familiar with the horrors of their persecution, of the death under torture of so many of their number, and of the confessions, recantations and final death by burning of their last Grand Master.

But the part of all this tale which would undoubtedly have been the most familiar to the young Portuguese Prince was the subsequent fate of the Order in his own country, where King Diniz was so conscious of the help that the Templars had afforded to his kingdom in the past, by helping to expel the Moors from the Algarve, that he modified the Pope's instruction to persecute the Order. He allowed the knights to escape and contented himself with occupying their property as a prelude to setting up his own national order – the Military Order of Christ – which in due course absorbed much of the Templars' wealth and property. The new knights continued to wear the Templars' crusading cross as their emblem and almost certainly frequented the court, introducing an aura of chivalric romance. It was small wonder they made a deep impression on the young Prince.

The early years of Prince Henry's life were never recorded in any detail, the chroniclers seeing their task as starting with those great events which were to be their preoccupation. But Prince Duarte's writings do throw some light on the affection between the brothers, on their enthusiasm for the knightly preoccupations of the court, and on the life they led. Predictably, horsemanship played a large part in this life; indeed Duarte wrote a separate treatise on horsemanship. All the Princes early acquired a mastery of equestrian pursuits, and the chief among these was hunting. Game was plentiful in the nearby Alentejo region which stretched eastward to the Castilian frontier and southward to the province of the Algarve. Over some of this, cork and olive trees had already been planted, but over most of it there was no productive crop and the region was one huge sporting forest, abounding in deer, wild boar, hares and – further north – wolves. Falconry was also a favourite pursuit and Prince Henry, together with his eldest brother Duarte, was not infrequently charged by the King with the agreeable task of organizing large-scale royal hunting parties. At these events considerable etiquette was observed and they were very different from the casual, marauding hunting expeditions of King João's youth, when the daughters of the local peasantry were as likely to be the object of the chase as the four-legged inhabitants of the forest.

When the uneasy truce with Castile was finally consolidated and formalized into a treaty of friendship and non-aggression in 1411, which was specified to last for a hundred and one years (as opposed to the perpetuity specified in the similar treaty with England), hunting was almost the only knightly sport open to the young Princes. Almost, but not quite. For there remained the joust. Elsewhere in Europe by the beginning of the fifteenth century, tournaments had reached a degree of extravagance – both in expenditure and behaviour – which risked rendering them absurd. More than a century before a ballad entitled *Des Trois Chevaliers et del Chainse* had recounted a typical instance of chivalry carried to ludicrous extremes: a courtly lady had sent her chemise to her three knightly admirers with a message to the effect that the one who truly loved her would wear it instead of his coat of mail. It seems unlikely he survived long to enjoy her favours. Such extravagant follies were the stuff of decadent tournaments. The young Portuguese princes performed in the lists with skill and courage, however, and their militaristic father was as proud of their prowess in this field as he was of their performance at the chase.

It was therefore perhaps hardly surprising that, with the Peace Treaty signed with Castile, the King should have decided on a tournament as the appropriate occasion on which to confer knighthood on his three eldest sons, Duarte, Pedro and Henry, who were respectively aged twenty-one, twenty and eighteen at the time. As became a King and Court who were so preoccupied with chivalric protocol, the tournament was to be no ordinary one, but a veritable epic of the joust. There were to be separate contests throughout a whole year. Invitations were to go out to all the courts of Europe to send noble competitors. The prizes and hospitality were to be in keeping with the splendour of the occasion and specially designed to impress the visitors. Indeed the chronicler quotes the King as declaring, 'I shall give presents of such magnificence, above all to the strangers, that the greatness and pleasantness of these gifts will oblige these lords to speak of them with admiration to all their friends.' He may well have had in mind something like the solid gold lances, swords or axes which were to be given as prizes by the Duke of Burgundy at a jubilee tournament a few years later. The pageantry was to be truly sumptuous. The young Princes would receive their knighthoods amidst these scenes of splendour. No project could have so well epitomized the spirit of the Court in which Henry grew up.

And yet it was not to be. The objection came from a quarter which – on the face of it – seemed unlikely. It was not from the King's treasurer, João Afonso de Alenquer, who might reasonably have raised objection to the expense. It was from the Princes themselves, whose concept of chivalry demanded a more positive way of winning their spurs. When the King outlined the projected tournament to them they respectfully begged leave to consider his generous proposition before accepting it.

Their considerations have been recorded by the chroniclers. They were attended by their older half-brother – the bastard Count of Barcelos – who had recently returned to the Court from his own foreign travels. Some subsequent commentators have suggested that the Princes' objections to the tournament were evidence of the Renaissance attitude of scorn towards such events, an attitude encapsulated in Petrarch's celebrated question, 'Where do we read that Cicero or Scipio jousted?' But there is nothing to substantiate this speculation, and it seems much more probable that the Princes' long familiarization with the deeds of the Crusaders – especially of the Templars – was responsible for their preferring a real military venture to a charade. We must wait longer for any convincing evidence of Renaissance thought processes in Prince Henry.

The chroniclers relate that it was, in fact, the King's treasurer who, having been taken into the Princes' confidence about their misgivings regarding their father's projected tournament, first suggested a suitable objective for a military adventure. The Princes themselves had recognized that Granada – the most obvious and convenient Infidel target – was a preserve of their Castilian cousins, and had been contemplating seeking service under some foreign Christian sovereign as a means of winning their spurs more honourably. The experiences of the Count of Barcelos indicated that various opportunities might be open to them. The treasurer now suggested to them that they had a natural objective near at hand, in which they might reasonably hope to engage the interest of their own sovereign and father – the conquest of the Moorish stronghold of Ceuta on the North African coast (almost opposite Gibraltar). The attraction of Ceuta as an objective was much enhanced by the provocative use the Moors were making of this port; it was from here that many of their piratical activities in the straits were launched, and it was to this town than many Christian captives were brought to start their long and humiliating years as Moorish slaves. It was a worthy objective for a national crusade.

Although the idea had already been propounded to the King, it was the urging of the Princes which fixed his mind upon the Ceuta enterprise. However, the King saw numerous objections to the project: the diminutive size of the Portuguese army and the lack of any fleet adequate to the task of transportation; the expense of the project and the undesirability of imposing further taxes on an impoverished people; the dangers of leaving his newly-established kingdom undefended; and, not least, the risk that by seizing Ceuta he would facilitate the Castilian conquest of Granada and thus upset the delicate balance of power in the Iberian Peninsula.

All these objections were answered by the Princes. The small standing army of the King – some 3,500 cavalry – could legitimately be augmented by exacting the levies of armed men which both the nobility and the chartered cities were obliged to provide; in addition, a network of master-bowmen throughout the country had the duty to recruit, train and lead companies of local crossbowmen. The construction of an adequate fleet was not beyond the capacity of so richly wooded a country. Additional taxes could be avoided if loans were raised from the emerging merchant class, and – of course – there would be substantial savings to the exchequer resulting from cancelling the marathon tournament. As for the security of the state, only Castile could menace the homeland's temporarily unprotected frontiers and there was no reason to doubt the good faith of the signatories of the 1411 Peace Treaty. The further objection, that the conquest of Ceuta would facilitate the Castilian objective of expelling the

Moors from Granada, was found to be unworthy: all Christendom should surely rejoice in the discomfiture of Islam and local considerations of regional balances of power should not be permitted to frustrate the holy purpose. The evidence suggests that these secular objections were disposed of with relative ease.

The main debate – according to the evidence of the number of meetings held – centred around the religious and chivalric implications of the project. The King consulted his confessor, who in turn consulted the more discreet theologians throughout the country; they unanimously declared that if the enterprise were undertaken for the greater glory of God then 'making war on the people of Africa' could not fail to be a holy task. When the project was broached with the Queen, she readily succumbed to the King's insistence that the Christian bloodshed of his early campaigns (not to mention the cold-blooded murder of the Count of Ourém) could only be assuaged by 'washing his hands in the blood of the Infidel'. The records show that Prince Henry was particularly adept at handling his father's anxieties on these scores, and it was to him and his brother Duarte that the King entrusted the arrangements for the final consultation regarding the project: a presentation of it to that ultimate arbitor of chivalry – the Constable Nun' Álvares. To avoid the publicity and speculation which would have followed summoning the Constable to Court, the Princes arranged a royal hunt near the Constable's estates at Montemor in the Alentejo; in these circumstances a meeting between the sovereign and his most revered subject took place with a minimum of ceremony. The Constable appears to have asked no questions regarding the practical aspects of the proposed venture; he unhesitatingly declared it a noble and inspired *geste*. He repeated his verdict shortly afterwards at Torres Vedras in front of the Council of State, whose approval was predictably enthusiastic. The Court had committed Portugal to a hazardous adventure which was more of a latter-day crusade than a rational campaign.

But much subsequent rationalization has taken place. Even Zurara, the Court chronicler, writing with hindsight, inclines to credit Prince Henry with the vision for seeing, from the inception of the idea, Ceuta as a gateway opening up Africa to Portugal. Subsequent Portuguese historians, such as Oliveira Martins, have interpreted Henry's silence, when confronted with a sand-table plan of the port and defences of Ceuta, as evidence that 'the first glimmerings of the idea formed in his mind which afterwards gave birth to the great nautical school at Sagres . . . and the commencement of all modern scientific navigation.' Such theories seem to predate the awakening in Prince Henry of more complex mental processes.

More recent Portuguese historians, like António Sergio and Jaime Cortesão, have attributed economic motives, such as the desire to acquire for Portugal the cereal crops of the Ceutan hinterland, to Prince Henry and the architects of the expedition. But there would have been easier ways of acquiring cereals than by attacking a reputedly impregnable Moorish citadel. The overwhelming volume of evidence about the way the Ceuta decision was taken, and about the mental climate in which it was taken, suggests that – unlike the other European Courts to which the chroniclers mistakenly attributed chivalric motivation – the Court of King João really was obsessed by the philosophy of chivalry to the almost total exclusion of more substantial and familiar motives, such as personal advancement and economic betterment. Later in Prince Henry's career the chroniclers' attribution of chivalric motivation is more suspect, becoming – in fact – a prime example of that capacity for self-deception which Huizinga so percipiently analysed. The stages by which Prince Henry parted company with this heroic image of himself emerge in his subsequent career. But for the moment the illusion and the reality were in harmony. When he embarked on the Ceuta enterprise, Prince Henry – possibly for the last and only significant time in his life – was following the dictates of his extraordinary, artificial and uniquely antiquated environment. He was an uncomplicated Crusader.

Chapter 3

The Model Crusade

'This is the happy warrior, this is he
Whom every man in arms should wish to be . . .'

Wordsworth

The whole Ceuta expedition, from its preparation to its conclusion, was an epic of medieval heroics. Even discounting the naturally chivalric embellishments of the chroniclers, the facts of the story itself read like a passage from Malory's *Morte d'Arthur* or Spenser's *Faerie Queene*.

Once committed to the project, King João entered into the spirit of the adventure like a young man. He devised a plan for a secret reconnaissance of the port and defences of Ceuta, despatching the Prior of the Order of the Knights Hospitaler on a mission, ostensibly to the widowed Queen of Sicily to discuss marriage prospects for one of the Princes, but in fact instructing him to put in at Ceuta on the outward and return journey to take soundings in the harbour and to study the ramparts. This the Prior accomplished without arousing any suspicions, returning to confuse the King and Court – who received him at the summer Palace of Cintra outside Lisbon – by declining to relate his findings until provided with two sacks of sand, a bag of beans, a roll of tape and a basin. The King feared that he was going to be subjected to a display of magic until the Prior, who withdrew to an inner room with his simple requirements, eventually confronted the King with a detailed sand-table model of the town, port, walls and beaches of Ceuta.

Preparations for the expedition occupied nearly two years. The

King wished to free himself of routine responsibilities in order to co-ordinate the work and he entrusted the administration of the realm to his eldest son, Prince Duarte. The Princes Pedro and Henry were despatched respectively to the Alentejo and to the North to recruit, equip and train forces for the expedition. In Henry's case his rôle included the task of building an additional fleet in his native city of Oporto, in which to transport his army southward to join the main invasion force at Lisbon. Prince Henry's powers were vague but far-reaching in that he held a commission from the King enjoining all subjects to obey him; he had no detailed instructions. His task was complicated by an outbreak of the plague in Oporto.

From the outset Prince Henry appears to have communicated his enthusiasm to the people of Oporto, no small feat for a nineteen-year-old youth when it is remembered that he was not empowered to explain the destination of the armada. It was not only the dockyards that worked throughout the days and nights, but also the armourers, the butchers slaughtering animals and salting meat, the tailors making tabards, the sailmakers, the rope-makers, the coopers and every craftsman in the city who had something to contribute to the task in hand. The citizens forbore to eat meat, which could be preserved for the expedition, and lived on offal – thus allegedly accounting for the name of *tripeiros* which is still applied to the inhabitants of Oporto. The city was temporarily imbued with that collective dedication which occasionally overtook medieval towns – notably Chartres during the construction of her cathedral – when some mighty work to the glory of God was afoot.

And there was no lack of spectacular incidents to sustain the dedication of the inhabitants. A monk at the São Domingos monastery, at vigil before dawn, claimed to have seen a vision of King João kneeling before the Virgin Mary, his hands extended to heaven whence a shining sword was being extended to him. Tremors of excitement went through Oporto as the story was circulated.

Hand in hand with religious ecstasy went chivalric zeal. A

ninety-year-old knight, Ayres Gonçalves de Figueiredo, rode to Prince Henry's headquarters, attired in his ancient coat of mail and at the head of his men-at-arms, to offer his services to the cause. Two old squires from Bayonne, who had preserved their weapons since the King's campaigns of two decades before, similarly presented themselves and were not to be deterred from further service. The call of chivalry had even reached more distant foreign courts. A French knight with a reputation for courtly literary achievements – Antoine de la Salle – and others from Picardy and Normandy presented themselves as volunteers. A German baron did the same, and a German grand duke expressed interest in bringing his retainers to swell the Portuguese forces. King Henry V, unable to resist such a romantic appeal, took the unusual step of authorizing the recruitment of several hundred archers in England. The Earl of Arundel offered to send a fleet. (This last offer was withdrawn after the resumption of the Hundred Years' War with France and may in any case have been partially motivated by the Earl's kinship to the Portuguese King.) That help should have come from so many and such varied quarters is doubly remarkable considering that the objective of the expedition was still a jealously guarded secret, so jealously guarded, in fact, that the German grand duke withdrew his offer in understandable pique when not taken into Portuguese confidence on this important point. While other wars were brewing elsewhere in Europe, there was no shortage of battlefields for ambitious or frustrated soldiery. So the decisive reason for volunteering to join King João and his sons in their venture could only be that the Portuguese Court was renowned for its chivalric motivation and that any military venture it undertook could be relied upon to be a worthy cause for which a knight should take up arms.

But even despite this sentiment – or possibly because of it – it was becoming increasingly difficult by 1414 to maintain the necessary secrecy regarding the destination of an expedition for which such widespread and belligerent preparations were seen to be in hand throughout Portugal. Neighbouring kingdoms were showing signs of alarm. Castile – the old enemy – was naturally the first to grow nervous, and it is a further measure of the honourable esteem in which King João was held that he was able to quieten their nervousness by his hospitable entertainment of their envoys. The King of Aragon was similarly reassured that Portugal had no intention of either backing the claims of the Count of Urgel to his throne or in any other way becoming his enemy. With the Moorish King of Granada, whose alarm was more vivid than that of other neighbours, the Portuguese were less affable and found it harder to convey reassurance. Granada – that last Infidel foothold in the Iberian peninsula – was so obvious a target for a crusading Christian monarch that its rulers failed to realize that the honour of their destruction was the prerogative of Castile; they resorted to sending envoys to Lisbon offering bribes to Queen Philippa, in the form of an opulent trousseau for her daughter, if she would use her influence with the King on their behalf. These Moorish offers gave great offence to the self-consciously upright Court and were rejected, Prince Duarte adding scornful remarks to the envoys about behaviour more becoming merchants than knights. Granada was unconvinced and started to fortify its coast afresh. Ceuta might well be expected to follow suit. And further afield also, anxieties as to Portuguese intentions were causing potential embarrassment with Naples, Normandy (as a grandson of the Count of Boulogne, King João might have established a claim) and Flanders. Clearly, secrecy about the Portuguese destination was no longer enough; subterfuge also was necessary.

Here King João's reputation for latter-day chivalry stood in his way. He was considered unlikely to launch an unprovoked attack upon another Christian country without due cause. Happily a plausible cause could be found: for some years Portuguese shipping had suffered from acts of piracy by Dutch ships. King João despatched an embassy to Count William of Holland to complain, in public audience, about these acts; the complaints were found unacceptable and a formal act of war was declared. But Count William was privy to the charade in which he was participating; in private he loaded the Portuguese ambassadors with rich gifts and sent them home to Portugal with messages of greeting. Ceuta was lulled into forgetting any suspicions.

In Oporto, Prince Henry's work drew to its triumphant conclusion in July 1415. He sailed from the mouth of the Douro at Oporto to the mouth of the Tagus at Lisbon to join forces with his father. His own new fleet consisted of twenty war galleys (the caravel – never primarily a warship – was not to appear for some years) all flying the Prince's new standard bearing his new motto *Talent de Bien Faire*; his newly recruited troops were in newly tailored uniforms; the vessels were freshly painted and their figureheads freshly gilded. Contemporary accounts described the scene as reminiscent of a pageant – and so it was.

But like all romantic courtly epics, the Ceuta expedition had an interlude of noble tragedy. Prince Henry's arrival at Lisbon coincided with the news that Queen Philippa was seriously ill. She had apparently contracted the plague at Sacavém, a low-lying village close to Lisbon to which the Court had moved when the plague became serious in the capital. She had ignored outbreaks of plague in Sacavém itself, preferring to stay at her devotions than move with the rest of the Court to Odivelas. When eventually she rejoined the King and the Princes, her pallor was attributed to fasting – against the advice of her physicians – in devout preparation for the hazards that were awaiting her husband and sons at Ceuta. Her condition quickly deteriorated, however, and her three elder sons were summoned to her bedside where a scene ensued which, according to the accounts related by the chroniclers, would have graced the pages of Ronsard. The Queen distributed fragments of the True Cross, to be worn by the Princes and their father; she presented each of the three Princes with a sword and injunctions to use it as became a true knight; she recalled to them the English proverb that while an arrow could easily be broken singly, a quiverful could be broken by none – and thus united her sons would be inviolable; she remarked that the wind was set fair for Ceuta and blessed the expedition. As she expired, any doubts that lingered in Prince Henry's mind about the sanctified nature of the crusade on which he was about to embark must have expired also.

After some debate, in which Prince Henry was among those who successfully persuaded the King and Council that the Queen's death – even following, as it did, an eclipse of the sun and a summer of plague – was not to be construed as a bad omen for the expedition, it was decreed that mourning was to be discarded and the pageant continued. The ships were once more decked with flags and Prince Henry declared his own force ready for immediate embarkation. When the whole fleet assembled it numbered two hundred and forty vessels, including twenty-seven war galleys with three tiers of oars, thirty-two with two tiers, sixty-three troop-transporting ships and over a hundred stores ships. This armada was propelled by thirty thousand sailors and oarsmen and carried twenty thousand soldiers. Additionally there was a sizable contingent of English archers (which is the more remarkable when it is remembered that 1415 was the year of Agincourt when Englishmen would 'hold their manhoods cheap' that were not there).

This was the force that sailed out of the estuary of the Tagus on the 23rd of July; three days later they rounded Cape St Vincent and Prince Henry saw (probably for the first time and certainly in the most memorable circumstances) the Sagres peninsula where he was to spend so great a part of his later life; the same night they anchored in Lagos harbour, at the western end of the Algarve coast. It was here that the destination of the expedition was revealed to the participants, significantly in the course of a sermon by Brother João Xira which incorporated the publication of a Papal Bull declaring the enterprise to be a crusade against the Infidel. The prelude to the campaign could hardly have been more auspicious from a chivalric point of view; the subsequent course of events was to exceed even these high expectations.

The voyage was punctuated with heroic incidents. The first of these occurred when Prince Duarte came aboard Prince Henry's ship to spend a fraternal evening and to stay aboard for the night.

Duarte was sleeping on deck and Henry below when an oil lantern overturned and started a fire in the ship. Duarte roused his brother and took him to the scene of the trouble, where Henry promptly picked up the burning lantern in his two hands and threw it overboard. The fire was subsequently extinguished without difficulty. Henry's only concern was as to whether his injured hands would heal sufficiently fast to enable him to wield his sword with a firm grip in the forthcoming encounter.

The next incident on the voyage reminds one of the gallantry of Sir Philip Sidney at the siege of Zutphen more than a century and a half later. The comptroller of Prince Henry's household – a certain Fernandalvares Cabral – was delirious with 'a touch of pestilence' and suffered from the illusion that his master was surrounded by Moors with no one to come to his rescue. Prince Henry's personal physician who was attending the patient reported to the Prince that, while if it were only the comptroller whose welfare he had to consider he would have suggested that his master calmed the man by his presence, as the Prince's medical advisor he had to insist that the Prince should keep well away from the man for fear of infection. Prince Henry overrode the advice and personally comforted the invalid, thereby deepening the affection in which his household held him.

The fleet was becalmed for a week before reaching first Tarifa, then under the charge of a Castilian governor, and later Algeciras. Here the Portuguese assisted in capturing and hanging a local bandit who had been terrorizing the villagers. Like knight errants, they performed shining deeds along their way.

But their obsessive chivalry was soon to lead them into one of those typical absurdities reminiscent of Henry V and his coat-armour. When the fleet crossed from Algeciras to Ceuta on the 12th of August, they ran into bad weather. They moved from anchorages on the east side of the Ceutan peninsula to supposedly safer moorings on the west side, where a severe gale compelled the King and a part of the fleet to retire to the shelter of Algeciras. The Constable – the elderly Nun' Álvares – was, however, in charge of another part of the fleet which remained in Ceuta. He rejected the advice both of the soldiers under his command, who wanted to land and fight, and of the sailors, who wanted to stand off and keep their vessels at a safe distance from the rocky shore. The military advice he rejected because it would have been improper for him to commence hostilities in the absence of his lord the King; the nautical advice he rejected because, having come within view of his enemy, he thought it ungallant to retreat. In this perilous state the Constable's ships drifted for two nights and a day. Happily for all concerned on the Portuguese side, the King then sent the old warrior an explicit order to withdraw and join him at Algeciras.

A week passed while the fleet regrouped and waited for the storms to abate. Although they did not know it at the time, the week was not entirely a lost one for the Portuguese, since the Moorish ruler of Ceuta – Sala-ben-Sala – foolishly assumed that the Portuguese armada had abandoned its intentions of attacking Ceuta and consequently dismissed the Berber reinforcements who had been summoned to swell the ranks of the defenders. On the 20th of August the fleet again crossed the straits under cover of darkness and made ready for a dawn assault on the beaches below the city walls.

Now that the moment for the attack had arrived, the chivalric protocol again came to the fore. Prince Henry had long before requested from his father the honour of being the first to land. His resolution during the last days persuaded the King to grant the request, particularly as the Prince's burnt hands appeared to have recovered. But at the last moment one of the household of Henry's bastard half-brother, the Count of Barcelos, snatched at this honour by leaping prematurely into the water. Prince Henry, considerably angered, thereafter led his followers ashore without awaiting further orders.

The battle which ensued was predictably depicted by the chroniclers as an unqualified tale of the highest heroism, and so it almost certainly must have appeared to Prince Henry and the other Portuguese commanders. Moorish skirmishers on the beaches, who included at least one Numidian giant, hurled rocks at the landing party, smashing the visors of their armour in some cases. The Portuguese drove them back and managed to follow the retreating defenders through the city gates – Prince Henry's standard-bearer being the second man to force such an entry. Zurara reports how one knight – Vasco Fernandes – felt it was less than heroic to enter by the same gate which the Princes had used, and went around the wall until he could find another gate 'which he straightway began to batter down'; having made his separate entry he was promptly overwhelmed and killed. (Doubtless Prince Henry was reminded of the forty Templars at Ascalon: this was how he had dreamt of war.) The King then landed with the main body of his force and the inner ramparts were also breached.

Almost all the *dramatis personae* of the chroniclers' earlier passages receive favourable mention in their accounts of the battle, including the King's treasurer and the ninety-year-old knight. There is no doubt that Prince Henry fought with special valour and effectiveness; although missing in the fray for two hours at one stage, he was subsequently discovered besieging a tower into which a much more numerous force of Moors had been driven to seek refuge. By afternoon, virtually the whole city was in the hands of the Portuguese and looting had taken over from serious fighting as the main occupation of the army. Large numbers of Moors lay dead in the streets and even larger numbers had fled into the surrounding woodlands.

The battle for Ceuta had something in common with another great Christian victory over the forces of Islam nearly five hundred years later – at Omdurman. In both battles, the European force displayed considerable dash and courage and was convinced of the glory of its own victory. But in both battles also the primitive nature of the Mohammedan arms and the very light casualties inflicted on the Europeans suggest that the contest was hardly an equal one. The Ceutan Moors, however viciously they might have hurled rocks at the landing Portuguese, had no weapons or armour comparable to those of the invaders and no answer to the English archers. And the fact that, after a day's fighting which left such piles of Moorish bodies that these had to be dumped in the sea, only eight of the Portuguese army had been killed (including one Englishman to whom the chronicler Zurara gives the curious name of Inequixius Dama), suggests a disparity in the contest. While the Portuguese were debating whether to assault the final citadel on the same evening or the following morning, it was found already to have been evacuated by the Moors. The victory could not have been more complete.

If the Moors were no match for the Portuguese militarily, it none the less became abundantly clear during the sacking of the city that their living standards compared very favourably with those of the invaders. The intruders stumbled, in their search for gold, silver and jewels, on many other valuables the worth of which few of them could appreciate: there were oriental carpets and hangings, marble terraces, mosaic floors; warehouses filled with wheat, rice and salt; huge jars of pepper, cinnamon, cloves, ginger and other spices. All these riches constituted the fruits of trade with Saharan Africa and the Indies, which flowed in to Ceuta through the caravan routes from the south and east. As a part of the Kingdom of Fez, Ceuta was in touch both with the interior of Africa and with the virile Mohammedan world of Asia Minor. These prizes did not pass unnoticed by Prince Henry.

But for the moment, the Prince was preoccupied with the glories of conquest. The principle mosque of Ceuta was rapidly prepared for consecration as a Christian church (the reverse process was to happen to Santa Sophia in Constantinople just thirty-eight years later); two great bells, stolen from a Sines church by Moorish pirates, were discovered and installed; the King declared that high mass would be celebrated on the Sunday following the capture of the city.

This occasion was, for Prince Henry, the climax and crown of his early life. Together with his brothers Duarte and Pedro he presented himself before the King, dressed in his battle-dented armour and carrying his dead mother's sword at his side. A solemn *Te Deum* was sung. Then the King summoned the Princes in turn and taking from each his sword dubbed them as knights.

Prince Henry kissed the blade of his sword before handing it to the King; the sacred memory of his chaste mother mingling, no doubt, in his mind with the solemn and honourable ceremony.

It was the vision of a dream come true. Never again was life to appear in such simple, primary colours. As Prince Henry's intellect developed, so the shades of thought clouded the course of future undertakings. The future was to be laden with questions, and questions not only of geography and science and navigation, but questions of motive and principle. In facing these questions, part of Prince Henry's mind was to remain hypnotized by the simple chivalry of the Court and the Ceuta expedition. The vision of a dream was to become a mirage – leading him into fatal enterprises and false conclusions. But this was in the future.

Chapter 4

First Enquiries and First Discoveries

> 'What should we do but sing His praise
> That led us through the watery maze
> Unto an isle so long unknown,
> And yet far kinder than our own?'
>
> Andrew Marvell

In the literal, as well as in the metaphorical sense, Prince Henry came of age in the year of the Ceuta expedition. What sort of a young man did he appear to his contemporaries? The answer is a formidable, rather than a lovable one.

Certainly physically he was formidable. He was not only tall and long-limbed but broad and large-framed. His black hair was thick and shaggy. His complexion was weatherbeaten and unusually dark for a Portuguese. His natural expression was an austere one and even his admirer Zurara admits that 'the expression of his face at first sight inspired fear in those unaccustomed to him'. Even had he not been a Prince of the Blood, no one would have found him a young man to be trifled with. Observers noted that his ambition was visible in the set of his jaw.

Appearances did not belie the inner man. While other twenty-one-year-old princes were womanizing or contemplating marriage, he appears to have had no time for female company. Nor, in a Court attended by numerous squires and with a household of his own abounding with young pages, was there ever any suggestion that the tendency of his affections was less orthodox. It was also at this time in his life that he permanently renounced wine and adopted a style of total abstinence. All his energies were to be concentrated on the task in hand.

And tasks there were. Immediately on return to Portugal from the Ceutan expedition, King João bestowed further honours on his sons: Prince Henry was created Duke of Viseu and Prince Pedro, Duke of Coimbra. (This was the first occasion on which the title of duke had been used in Portugal. It was to prove unfortunate that the bastard Count of Barcelos was not honoured like his half-brothers, for by the time that the Dukedom of Braganza was to be bestowed on that side of the family, its sense of bitterness and grievance was already fatally established.) The new titles carried fresh responsibilities and Prince Henry was sent to live for a while at Viseu, in central Portugal, to concern himself with the security of the neighbouring frontier with Castile.

The assignment was short-lived. King João had left the conquered city of Ceuta defended by a garrison under the command of the Count of Viana. Within three years the Count found himself threatened by an alliance of the Moors from the Kingdom of Fez, in the interior of Morocco, and the Moors from the Kingdom of Granada, on the opposite side of the straits of Gibraltar. He called for reinforcements from Portugal and the King despatched Prince Henry in command of a relief expedition. The appearance of his fleet was sufficient to throw the Moors into disarray; they fled and rapidly concluded an unconvincing truce.

Prince Henry stayed on for several months in Ceuta – months that were to have an immensely formative effect on his future. He had long talks with the Count of Viana. The latter explained the real predicament of Ceuta: the city was a terminal for the African trading caravans which came both along the Libyan coast from Egypt and Baghdad, and across the Sahara from the fabled Sudan (the area between Senegal and Niger and not to be confused with the Sudan of the upper Nile) and Timbuktu. At the time of its conquest Ceuta contained 24,000 commercial establishments (though many of those may have been no more than market stalls) dealing in gold, silver, copper and brass as well as in silks, spices and weapons, imported from the orient and the interior of Africa. But – as the Count explained – since the Portuguese had seized the city, these supplies had dried up. Ceuta was an empty market place. The only ways to revitalize the city were either to establish a lasting peace with its Moorish neighbours (an unthinkable solution for a crusading prince) or to conquer the interior and provide the city with an agricultural hinterland.

It was in the contemplation of the latter alternative that Prince Henry began to collect information about what lay inland. His imagination was quickly fired with stories of much more than agricultural interest; the cereal crops of the coastal plains were the least of the attractions of the interior. Of much greater appeal to him were the more distant products which reached the coast only after weeks or months of travelling by camel caravan. The camel had been introduced to the Sahara from its original habitat on the upper reaches of the Nile about seven centuries earlier, prior to which the bullock had been the normal desert beast of burden; the camel's potential for long-range trading had been systematically developed over the centuries. Of all the products which it transported, the most valuable and the most inspiring was undoubtedly gold.

From the merchants of Ceuta, Prince Henry learned of 'the silent trade'. From southern Morocco the caravans traversed the Atlas mountains and set out for the twenty-day ride southwards towards Senegal. They carried loads of salt and beads, many of the latter made from coral in Ceuta. When they approached the area of the Senegal River they beat loud drums to announce to the hidden and apprehensive inhabitants of the area that trade was about to commence. They then set out on the river bank piles of salt and cheap manufactured goods and withdrew. As soon as they were safely out of sight, the natives of the area – who lived in the open-cast mines from which they dug their gold – approached and laid a heap of gold beside each pile of foreign merchandise. Then they, in turn, withdrew. When the merchants again approached they either took the gold or, if they considered it insufficient, reduced the size of their own pile of merchandise and again withdrew so that the process might be repeated. Eventually further drumming signified the close of negotiations. Thus the silent traders never met and only rarely were the rules of the game broken by over-greedy traders.

All this had been a closed book to the Portuguese. Apart from a small community of privileged Christian merchants in Marrakesh, no Christians had been tolerated in the interior of Moslem North Africa. But at Ceuta, with its Moorish past, tales from the interior were commonplace. And better than tales, there were even maps. Most of these were the work of Jews from Majorca who had been allowed far greater freedom of movement than the Christians. Their maps were evocative rather than informative documents, marking passes through the Atlas, routes across the desert and walled cities, with more imagination than accuracy. But where total ignorance had prevailed before, half-truth had the force of evidence.

The addition of these new dimensions to his knowledge opened Prince Henry's mind to fresh horizons. His curiosity was aroused by matters that lay wholly beyond those military and courtly preoccupations that had hitherto dominated his thinking. He began to see the African interior not merely as a field of battle with the Infidel, but as a land of mystery and of opportunity. A thirst for knowledge was added to his existing thirst for glory.

But – for the present at least – his old sentiments predominated. Prince Henry was spoiling for a fight; in vain he lingered on in Ceuta hoping that the Moors would launch another attack. When the likelihood of this appeared to be receding, Prince Henry dreamt up other belligerent schemes which he discussed with his Council at Ceuta. The project which he favoured most was a surprise attack on the Moorish garrison at Gibraltar. The Council were almost unanimous in their opposition to so foolhardy a scheme. But – with the stubbornness and self-assurance for which he was already becoming celebrated – he overbore the Council and set sail with his fleet. What advice could not achieve, inclement weather did: a storm arose and forced the ships north-eastward towards Cape de Gata. They sheltered there for a fortnight and returned to Ceuta to work themselves up once more for the assault. It was not to be. A firmly worded letter from King João awaited Prince Henry, explicitly forbidding him to undertake fresh ventures and ordering him to return immediately to

Lisbon. The headstrong crusader was no longer to have his way.

When Prince Henry returned to Portugal, he acted very strangely. He did not settle at the Court and participate with his father and eldest brother in the management of the Kingdom. He did not return to Viseu to resume charge of a frontier district. He did not follow the example of his other older brother, Pedro, and go abroad on a tour of other Courts of Europe and thus broaden his outlook. Instead, he rode from Lisbon southward to the Algarve and did not finish his journey until he had reached the very south-western tip of Europe – that same promontory of Sagres which he had viewed for the first time on the voyage to Ceuta.

He could have chosen no bleaker spot. The cliffs of Sagres rise sheer from the Atlantic; gales sweep the tufts of rough grass which are the only vegetation; no roads stretched beyond the barren hamlet of Raposeira. Here Prince Henry hired a simple lodging and devoted his days and nights to contemplating the ocean. Although the chroniclers were not so disrespectful as to suggest it, many of the Prince's associates must have ascribed this curious conduct to pique at being prevented from attacking Gibraltar. It seems more than likely that this was an element in his behaviour. But there was much more than the pique of a frustrated soldier; there was also the curiosity of an awakened explorer.

For now that Henry had returned from Morocco, his head was full of those notions of commerce and enquiry which had been awakened in him by the Count of Viana and the merchants of Ceuta. He had been denied the opportunity to make an overland expedition to the interior of Africa: he would attempt an encircling one, by sea. This is not the point for a detailed analysis of Prince Henry's motives, for the simple reason that they were not yet fully developed in his own mind. For the moment, the immediate objective was enough. He commissioned the equipping of several square-rigged *barcas* for an ocean voyage – an act which was within his own competence and for which he did not require to seek royal approval – and he despatched them to the

Atlantic coast of Morocco, to report on the lie of the land. The only remarkable fact they had to report on their return was that they had lost one of their number: a gale had arisen and one *barca* had been swept westwards into the 'sea of darkness'.

Eventually, the missing ship returned to Sagres. Its captain had an exciting tale to tell. Having been swept some four hundred miles south-westward into the uncharted Atlantic, they had made a landfall at an unknown island with a sandy bay and a safe anchorage. The island was just six miles long and three miles wide. In gratitude for their deliverance, they named it Porto Santo. With their fresh water supplies replenished, the *barca* had negotiated a safe return to Portugal.

Whatever else Prince Henry may have had in mind, discovering Atlantic islands had clearly been no part of his plan. But he was not slow to recognize his good fortune. That his first expedition had resulted in a positive discovery (even if this was of a quite different nature from that which he had sought) gave him an opportunity to use this fact as a justification for further expeditions. At least, it would do so if he could show some benefit from his discovery. Doubtless it was with this in mind that he sent the two squires of his household who had made the discovery – João Gonçalves Zarco and Tristão vaz Teixeira – back to Porto Santo to cultivate the island. The third captain whom he sent to accompany these two was Perestrello (later to become the father-in-law of Christopher Columbus) and it was a misjudgement by Perestrello which brought the whole colonizing venture to disaster. He brought a pregnant doe rabbit with him to Porto Santo, which bred so fast that the existing vegetation and the imported crops were soon devoured by a horde of rabbits on an island devoid of any other animal or bird life which might have controlled their spread. The food gave out and the settlers quarrelled among themselves.

The attempted colonization of Porto Santo was thus not a success. It is even doubtful whether the island was a genuine discovery since it appears to be identical with one marked on a

Genoese map of 1351. But it led on to a more important development. Contemporary accounts vary as to how the neighbouring island of Madeira first came to Prince Henry's notice. One theory is that Zarco and Vaz Teixeira reported to the Prince that they had observed a low-lying cloud persistently on the horizon, which their sailors attributed to those 'vapours' they had been led by superstition to expect on the outer fringes of the 'sea of darkness'. Prince Henry sent them back to investigate and Madeira was revealed.

Another theory, which may well be complementary to the above, is that Prince Henry was already aware of the existence of the island. The reason he could have been was that one of Zarco's sailors claimed that when he had been a prisoner of the Moors he had met a fellow prisoner who was an Englishman and who told him a truly remarkable and romantic tale. This English sailor had been one of the crew of a Bristol ship chartered for an elopement by Robert Machin, an English squire who had fallen in love with and carried off a married lady of influential family called Anne d'Arfet. They had planned to sail to France but their ship – like Zarco's own – had been carried by gales out into the Atlantic, where eventually they had made a landfall on a fertile, wooded island which was larger than Porto Santo and which sounded much richer. Anne d'Arfet had died of exposure or distress or both, and her lover died of grief soon after. The sailors buried them by a home-made altar, set out in the ship's boat and duly reached the coast of Morocco where the Moors imprisoned them as Infidels. If Prince Henry had heard this tale, it must have added to his curiosity about those clouds on the horizon.

It seems likely that it was in 1420 that Zarco first landed on Madeira. He gave it that name because it was so densely wooded. Those very woods were the undoing of the first settlements. Zarco, being frustrated by the difficulty of clearing the ground, gave orders to fire the undergrowth. Very soon the resulting holocaust was completely out of hand. The whole island appeared to be on fire and some authorities maintain that it was seven years before the blaze was finally extinguished. A set-back this certainly was; a long-term disaster it was not, because the potash resulting from the conflagration was to prove an excellent fertilizer.

Prince Henry did not visit Madeira, but he took an active and immediate interest in it. Those organizing abilities which he had first demonstrated while building and provisioning his fleet at Oporto, before the Ceuta expedition, were now directed towards his new-found possession. He divided the administration of the island between Zarco and Vaz Teixeira, and he obtained his father's permission for the former to recruit settlers from the prisons of Lisbon. He demanded samples of the timber and even of the soil. He imported sugar cane from Sicily and vines from Crete and Cyprus. He sent cattle, seeds, tools and more men. He extracted from the island hardwoods for shipbuilding and domestic architecture in Portugal, as well as honey, fruit, vegetables, an excellent sugar crop and a distinctive wine which was shortly to become famous throughout Europe.

In a few years, Prince Henry* had established a prosperous colony of a different order altogether from the struggling settlements which the Castilian crown had set up in the Canary Islands. He received a good revenue from it which was to help him in future enterprises.

In the ten years between the conquest of Ceuta and the establishment of the colony on Madeira – that is between 1415 and 1425 – Prince Henry's life had taken on a new dimension. He was no longer an uncomplicated crusader. He had become aware of some of the complexities of life: of the fact that Christendom had no monopoly of civilization; of the fact that trade could be as great an incentive for endeavour as the pursuit of glory; of the fact that much that was unknown in the world was not unknowable. He had also learnt much about himself: that men could be made to do his bidding not only because he outranked them, but because his will was usually stronger than theirs; that he had a capacity for organizing the activities of others even when he was not leading them in person; that his mind was a restless one and

*The official responsibility was, of course, with the Portuguese Crown, and this has led some historians to question Prince Henry's primacy in the affair.

that he would not be content with answers that satisfied other men. But this knowledge of the world and knowledge of himself left him far short of equilibrium: his own priorities were not yet established. At thirty-one he was less sure of his destination in life than he had been at twenty-one. Yet his resolution was about to harden and the next decade was to be the one in which he made a mark on the world which successive centuries were not to eradicate.

Chapter 5

Into the Unknown

'We were the first that ever burst
Into that silent sea.'

Samuel Taylor Coleridge

The rediscovery of Madeira was an encouragement to further Atlantic exploration. Although some recent historians have challenged Prince Henry's prime rôle in this process, tradition and early accounts attribute to his persistence not only the finding of Porto Santo and Madeira, but also of the Azores. There were hazy reports, dating from the previous century, which would have been available to Prince Henry about islands further out in the Atlantic Ocean; but it was not until a knight and sea captain 'of illustrious family' called Gonçalo Velho Cabral reported the sighting of some rocks – whose formation reminded him of ants and provoked him to call them the Formigas – that there is any substantial record. (The earliest map to show the Azores in approximately the correct place is dated 1439 and marks them as being discovered in 1427, but there is some evidence to suggest that 1432 would be a more accurate date for the first landfall.)

The story of the discovery of the more substantial islands in the Azores has much in common with the story of the discovery of Porto Santo and Madeira. Just as Prince Henry had sent Zarco back, so he now sent Cabral back to look beyond his ant-like rocks; the sea captain lighted on the island of Santa Maria and found it almost as attractive and promising as Madeira. Cabral was granted the title of Captain Donatory of the island and spent three years in Portugal recruiting potential settlers; he returned with a colony which included 'men of rank and fortune' contrasting with the ex-convicts who had constituted a large part of the original population of Madeira. In other respects the development of the islands followed similar lines and a prosperous community was established in Santa Maria.

The discovery of the neighbouring island of São Miguel is alleged to date from a much later period in Prince Henry's life, but it seems appropriate to record it now. A negro slave is reported to have escaped from his master on Santa Maria and taken refuge on the wild mountain peaks; from there he spied another island to the north-west and, knowing the Portuguese enthusiasm for new discoveries, he returned to his master in the confidence that his news would earn his pardon. (Neither history nor legend relates whether it did.) The intelligence of the new island eventually reached Prince Henry who sent Cabral to investigate. It seems likely that the hazy earlier reports of islands, referred to above, may have indicated the likelihood of further land in this direction. Cabral returned, however, to report that no new islands were discernible. It was then that Prince Henry, with his characteristic blend of self-confidence and determination, persuaded Cabral that in fact he had sailed between Santa Maria and the unlocated island, and that he should return and try again. This he did successfully, giving the new island the name of São Miguel and the whole group the name of Azores – or hawks, because he identified these birds in the vicinity. (Some early accounts place these events between 1440 and 1444; others claim that the whole group – except the islands of Flores and Corvo – were revealed by 1439.)

Zurara suggested in his chronicle that the name São Miguel was chosen since this was the favourite saint of Prince Pedro and this prince had throughout taken an active interest in the Azores and in Madeira. Indeed, this has been quoted as contributory evidence that the credit for these discoveries does not lie with Prince Henry. Evidence indeed is short, but the established version of events is so consistent with Prince Henry's interest and character that it cannot lightly be laid aside. In any case, by the time the 400-mile-long chain of the Azores had been discovered, Prince Henry was

Ceuta a century after the Portuguese attack

Portrait of Prince Henry from Zurara's chronicle

beyond dispute established as – in Professor Parry's words – 'the royal specialist in overseas activity'. As with his other discoveries, Prince Henry ensured that the islands were properly colonized, that they provided him with an income and that they were under the spiritual jurisdiction of the Order of Christ.

While the discovery of such Atlantic islands was gratifying and their colonization profitable, Prince Henry's abiding purpose was not 'to sail beyond the sunset, and the baths of all the Western stars'. His main curiosity and ambitions lay not westwards but southwards. Morocco had fascinated him and further south fascinated him still more; if he could not strike out from Ceuta across the desert, then perhaps he could outflank the desert from the sea?

Throughout the decade following the discovery of Madeira, his main energies were directed southwards and here we are dealing with somewhat less conjectural and better documented events. Between the years 1425 and 1434 Prince Henry repeatedly and unsuccessfully sent expeditions down the west coast of Africa with instructions to break new ground – that was, to sail beyond Cape Bojador, the limit of existing knowledge. Such persistence could not be attributed to mere curiosity. His motives fascinated even his contemporaries, and Zurara attempted an analysis of them. He defined six separate reasons for Prince Henry's endeavours: to discover the unknown; to establish trading relations in a new area; to reconnoitre the power of the Infidel; to make contact with any Christian Kingdom (he can only have been thinking of Prester John's) which might be revealed; to convert the heathen; and lastly, to fulfil the destiny foretold in his horoscope. Three of those motives – those relating to religion – are completely compatible with his earlier career and are consistent not only with Prince Henry's own writings (for example, the letter to his brother quoted in the next chapter) but with the comments of one of his least religiously inclined captains – Cadamosto – who noted religious zeal as a principal motivation. One of the motives – the fulfilment of his horoscope – was a wholly medieval concept common to the calculations of most of his courtly contemporaries.

As regards commercial gain, there can be no doubt – as we shall see in later chapters – that the pursuit of gold, ivory and slaves played a considerable part in the later voyages and may well, particularly remembering Prince Henry's interest in the riches of Ceuta, have been a factor from the beginning.

But neither the pursuit of converts to Christianity at the expense of Islam, nor obedience to a horoscope, nor the pursuit of wealth can account for the particular form of Prince Henry's endeavour. There were more promising fields for Christian mission work than the coastline of the supposedly empty Sahara deserts; the vague predictions of the horoscope could be 'fulfilled' in a score of more conventional ways; the acquisition of wealth could better be pursued by collaborating with recognized traders from the East, by piracy against the Moors or by exploiting his estates and positions in Portugal than by sending out fifteen fruitless expeditions down a barren and dangerous coastline.

One is forced to the conclusion that the first of Zurara's reasons – the discovery of the unknown – was the mainspring of Prince Henry's initial endeavours to round Cape Bojador. This fact itself gives clear evidence of a very different frame of mind than the conventional chivalric one which had hitherto been dominant in his character. And the way in which he set about achieving his objective provides additional evidence: superstitions were rejected; the study of astronomy and other relevant sciences was related to clear objectives and not pursued – as so often in the medieval world – as an esoteric metaphysical exercise; empirical methods were employed and experiment made the basis of further experiment. It was a Renaissance prince, not a medieval knight, who achieved the rounding of Cape Bojador.

The sheer weight of the superstitions with which he was confronted was formidable indeed. Although Ptolemy's geographical theories were generally accepted – including the fact that the world was round – by educated men, for the uneducated the unknown still held mysterious terrors. Many seamen believed that as the southern hemisphere was approached the sea began to boil. Others believed that it turned into slime, inhabited by grotesque monsters. There were even those who still believed that anyone sailing too far to the south, or to the west, risked sailing off the edge of the world. Others again* asserted that there were mountains on the African coasts that 'attract men to them as a lodestone attracts iron, and that men laugh while being attracted, and at last are held fast.' A still more widely held belief was a variant of this, to the effect that there were rocks of such magnetic force that they drew all the metal bolts and fittings out of the timbers of ships and that the latter fell apart in the forbidden waters of the south.

There was a good deal of circumstantial evidence to reinforce these terrifying tales. For one thing, although ships had been known to attempt to sail into the unknown southern seas, none had returned. The limit of exploration was Cape Bojador, which protruded into the Atlantic a thousand miles south-west of Tangier and a hundred miles south of the Canary Islands. And this Cape had some very bizarre characteristics which seemed to indicate to the simple-minded that the Almighty did not intend them to go further. The cliffs of the Cape crumbled into the sea sending cascades of red sand and spray into the air. Shoals of silvery sardines agitated the surface of the water. Waves crashed over reefs. Currents met in whirlpools. The blinding sunlight on the uninhabited desert coastline gave no indication that life could be supported. Those looking for symptoms of the end of the world – or at least of the end of the navigable waters of the world – had some reasons for thinking they had found them here.

There were also severe technical difficulties to add to the psychological ones for those who battled their way down the inhospitable African coastline. The shallows near to the coast were a hazard to those who hugged the shore, and the risks of being swept out into the uncharted Atlantic were a deterrent for those venturing further from land. Adverse winds and currents bedevilled the return voyage, and were thought (even by more enlightened sea captains) to be insurmountable south of Cape

*Pietro D'Abano's *Conciliator Differentiarum*, c. 1342.

Bojador. Even if a return journey were effected, there were always Moorish privateers lying in wait for exhausted mariners in the more frequented waters off the Barbary coast and many a Christian crew had been overwhelmed and sold into slavery.

Prince Henry was determined that these problems should not prevent his captains from rounding the crucial cape. During the years between 1424 and 1434 it is calculated that he sent not less then fifteen expeditions down the African coast with this objective. Always his captains returned to him with explanations of the impossibility of the task. Prince Henry rewarded them for their endeavours but did not accept their explanations. In 1433 he sent a determined squire from his own household – Gil Eannes – on the familiar and seemingly hopeless task. Eannes returned with the familiar excuses for failure and, according to the chroniclers, Henry chided him with behaving like a cautious pilot from the Flanders coast who did not know how to use compass or chart. Eannes was sent back in 1434 and told not to allow himself to be diverted towards piracy or other objectives but to concentrate solely on passing Cape Bojador, in which case he would earn from Prince Henry 'both honour and reward'.

Gil Eannes sailed in a *barca* (not yet a caravel) of about 30 tons. This was a square-rigged and probably two-masted vessel, with an only partially covered deck space and a crew of about twenty. It did not have the very high poop, with more comfortable stern quarters, which was to be introduced later in the century. Nor was it as large as many of the *barcas* which the Portuguese had sailed during the previous century, when some of them, displacing up to 100 tons, had been used in the maritime trade with Spain, Normandy, Holland and England; but even when these larger *barcas* had existed, such vessels had been comparative rarities and specially mentioned as such in the documents (such as a shipbuilder's concession of 1377) of the time. By the time of Prince Henry, Zurara is making casual reference to 'a large *barca* of 30 tons and two small ones', suggesting that the more sizable vessels of the previous century were no longer the norm.

Living conditions on such ships were rough. Privacy was non-

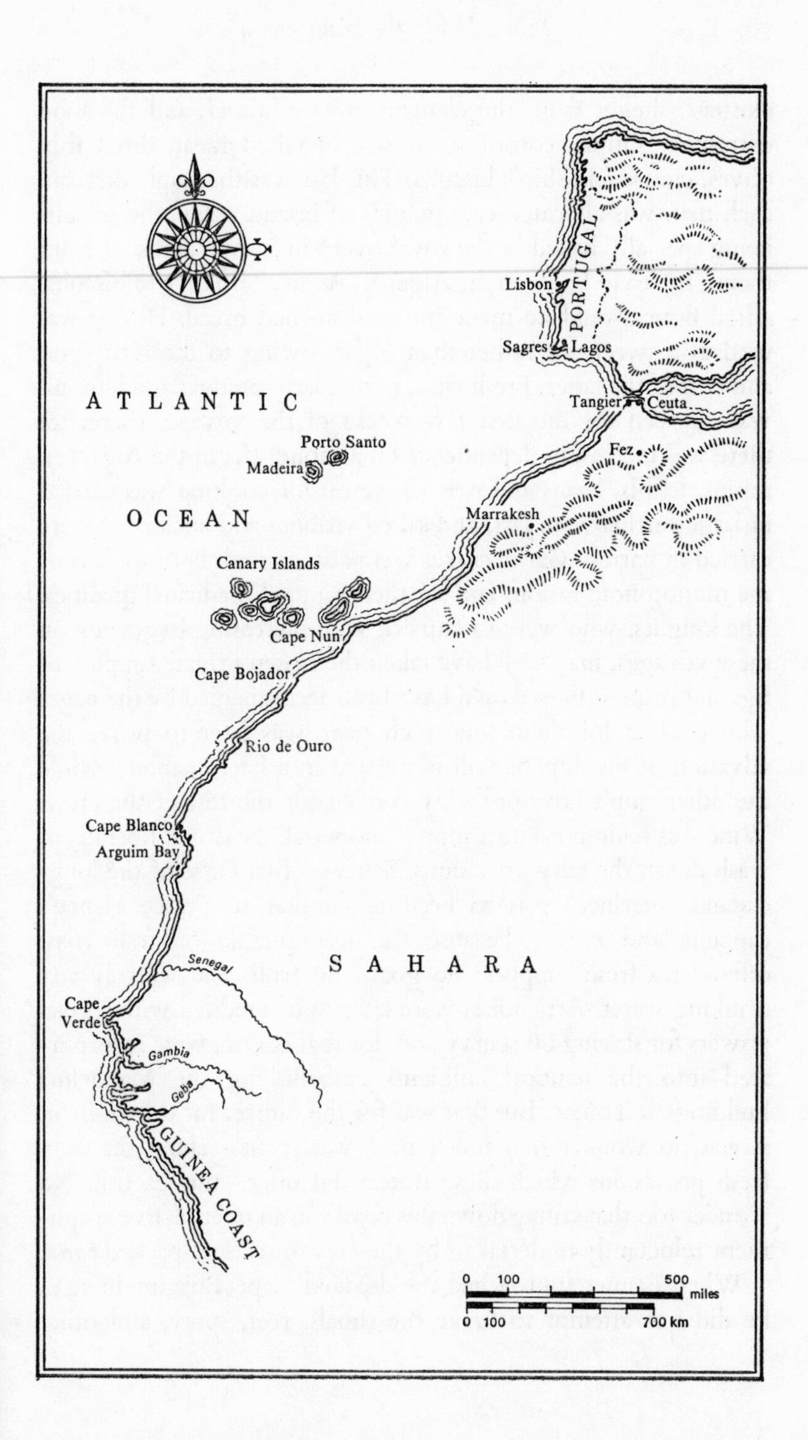
ATLANTIC
OCEAN
Porto Santo
Madeira
Canary Islands
Cape Nun
Cape Bojador
Rio de Ouro
Cape Blanco
Arguim Bay
Cape Verde
Senegal
Gambia
Geba
GUINEA COAST
SAHARA
PORTUGAL
Lisbon
Sagres
Lagos
Tangier
Ceuta
Fez
Marrakesh
0 100 500 miles
0 100 700 km

existent, shelter from the elements was minimal, and the food extremely simple, consisting mostly of salted meat, dried fish, olives, cheese and ship's biscuits. This last was the staple diet and each man was allocated two pounds of biscuit a day, the biscuits being specially baked in the royal ovens in Lisbon and sent from thence by sea to Lagos in the Algarve. As an alternative to biscuits, salted flour would be made into unleavened bread. Honey was used as a sweetener rather than sugar, owing to the rarity and and cost of the latter. Fresh fruit, particularly oranges and lemons, was enjoyed for the first few weeks of the voyage, thereafter there was increasing dependence on almonds (from the Algarve), raisins, lentils, beans and rice. Olive oil for cooking was carried in large earthenware jars, and salted sardines and anchovies were carried in barrels. Garlic cloves were also carried, both to flavour the monotonous rations and for their reputed medicinal qualities. The knights, who were to embark with increasing frequency on these voyages, may well have taken their own private supplies of figs and prunes; they would have been accompanied by the pages who cooked for them (one such page was later to prove the salvation of his ship, as will be related in a later chapter), while the other ship's boys probably cooked for the rest of the crew. Wine was seldom in short supply and would be drunk liberally to wash down the salty provisions. Scurvy – that curse of the long-distance mariner – was to become familiar to Prince Henry's captains and crews, because the inhospitable Saharan coast offered no fresh supplies: no goats, no fruit, and scarcely any drinking water. Artichokes were later to be credited with special powers for staving off scurvy and, for that reason, were incorporated into the nautical emblems embellishing the Manueline buildings at Tomar. But that was for the future; for the moment it was no wonder that much time was spent seeking the only fresh provisions which these waters did offer – the sea fish. No wonder too that sailing down this coast was an unattractive assignment reluctantly undertaken by the crew of a close-packed *barca*.

When Eannes approached the dreaded Cape Bojador in 1434 he did not attempt to brave the shoals, reef, spray, and other features that assumed such alarming guise there. Instead he changed course and sailing out into the Atlantic gave the Cape a wide berth; when eventually he altered course again and once more sighted land, Bojador – with all its alleged horrors – was behind him. He sailed on only a short way before anchoring and going ashore in a boat. There was no sign of human life on the scorched and bare Saharan sands, but he collected a few wind-swept plants of a type known in Portugal as St Mary's Roses and took these slender offerings back to Prince Henry.

He received a hero's welcome at Sagres. Prince Henry recognized that an invisible barrier of superstition and fear had been breached and that the way was open to progressive exploration down the African coast – possibly towards unconverted Infidels, possibly towards gold and ivory, possibly also towards Prester John's kingdom and the Indies themselves, but certainly – and above all – towards knowledge of the unknown. The feat of navigation by Gil Eannes had scarcely been remarkable, but the feat of will-power and determination by Prince Henry had been phenomenal.

Prince Henry wasted no time in following up this new development. The following year he sent Eannes back in his *barca* to go further, and he sent a second vessel – a *barinel* or oared galley – with him, under the command of another member of his household, his cup-bearer Afonso Gonçalves Baldaia. The *barinel* also had two masts, both square-rigged; as in the *barca* there would have been a forecastle and aft quarter at least partially decked-in, but only a make-shift awning over the midships where the oarsmen sat. Happily, at least on the outward passage, the main propulsion would have been the sails, as there would have been places for little more than a dozen oarsmen and the effort of driving the boat by man-power alone must have been heavy indeed.

Neither vessel was well suited to the voyage but, with the legendary fears of Bojador now allayed, they managed to reach a point on the coast some hundred and fifty miles further south than any previously explored. Once again a landing party was sent ashore. This time, tracks of men and of camels were found. They named the place *Angra dos Ruivos* on account of the gurnard, or gurnets, which they fished there from the sea, and then returned to Sagres for fresh orders.

The orders were predictable: to go back again and this time to try to establish contact with the unseen inhabitants. When Baldaia set out the following year with his single galley he had specific orders from Prince Henry to try to bring back an inhabitant of these southern deserts for inspection and examination at Sagres. To assist in the chase after prisoners, Baldaia embarked two horses on his ship, which must have considerably encumbered the restricted space amidships.

Baldaia succeeded in sailing nearly two hundred miles further down the coast than on the previous expedition, crossed the Tropic of Cancer and anchored in what he took to be the mouth of a huge river. In the hope that it might prove to be the Senegal River on whose banks the famous 'silent trade' in gold was supposed to take place, they called it *Rio de Ouro*; in fact, it was not a river at all, but a large inlet.

They wasted no time in following Prince Henry's injunction to try to locate and bring back specimens of the local inhabitants. It does not appear to have occurred to anyone that this might have been done other than by a marauding excursion to capture or kidnap one of the natives. Indeed, it is difficult to imagine Baldaia succeeding in negotiating with the Berbers – for it was they, and not negroes, who inhabited this shore – for one of them to join him voluntarily for the return voyage to Portugal; that such persuasion was not always impossible was to be demonstrated by others later, however, notably by Captain Robert Fitzroy who persuaded four Fuegans* to accompany him in the *Beagle* on the return voyage from Tierra del Fuega to England in 1830. But no such parleying was attempted by Baldaia and the

*Two of them, christened respectively York Minster and Fuega Basket, had a considerable impact on George IV's London and were a social phenomenon. Captain Fitzroy was as good as his word and returned them to Tierra del Fuego on his subsequent trip, but – understandably – they did not reintegrate happily.

first contact between the Christian explorer-missionaries and the indigenous population was to be a warlike one.

Two young squires of noble extraction were despatched inland on the two horses which had been brought for the purpose. They armed themselves with lances but did not take armour, doubtless reckoning that swift retreat would be their best defence. They followed a small river for some miles into the desert and then came on a party of nineteen natives, who were carrying spears, presumably for hunting. The natives, no doubt alarmed at the strange appearance of fair-complexioned youths on horses, and rightly fearing belligerent intentions, retreated to a rocky hillock and threw stones and spears at the intruders. 'They fought until evening', according to Zurara, who – as we shall see in later chapters – invariably described such encounters in epic terms, and several of the natives were injured while one of the Portuguese received a wound in the foot.

The squires eventually withdrew and rejoined their captain at the coast. Baldaia took a larger party the following day in a boat up the stream, while the 'cavalry' followed on the bank. He was determined to capture his living specimen if he could. But there was no sign of the previous day's party, except for blood on the rocks and sand, and some abandoned primitive possessions. These were gathered up and Baldaia returned to his ship and sailed another hundred and fifty miles down the coast, hoping to be more fortunate in his marauding further south. But apart from some curiously constructed fishing nets of woven bark, there were no fresh signs of human habitation.

Baldaia sailed for home, his ship laden with the skins of sea-lions which they had found in profusion in the newly discovered waters, and carrying the nets and chattels of the elusive natives. They had no one for Prince Henry to question; they did not even know if the men they had fought were Moslems or heathen; but they had at least seen human life in the unknown desert and – above all – they had sailed further down the coast than ever before. Prince Henry, as always, was benevolent and generous to such achievement; but he himself remained on dry land.

It often comes as a surprise, particularly to English-speaking students who know Prince Henry by the appellation of 'The Navigator', to learn that he never took part in any of his own expeditions of discovery. The question why this was so presents itself perhaps most forcefully in the context of the years which have just been reviewed, when successive expeditions failed to round Cape Bojador through the lack of will-power, authority or courage of their commanders. How was it that the man who possessed these qualities, and who above all others wanted to achieve the objective, did not himself take command? The question also arises forcefully at a later stage, when a regular stream of ships was going out, when a fort and trading station had already been established on the African coast, and when the ships were sailing in small fleets providing the security and amenities of a convoy.

It has been suggested by some recent scholars that it would have been as unthinkable for Prince Henry to go on one of his own expeditions as it would for the President of the United States to have gone in one of the early space capsules. The tasks of organizing the funds, collating the knowledge, assembling the experts and providing the incentive – it is argued – are quite separate from the tasks imposed in leading an expedition; the former is the function of the statesman, the latter of the sea captain. This argument would be more convincing if Prince Henry had been King and had been encumbered with the responsibilities – including survival – associated with the government of a realm; but he was not, and indeed he played only a very fitful part in public life. Nor had he other obligations, of a dynastic, administrative or family nature such as would have precluded his absenting himself from Portugal for months or even – had disaster overtaken him – permanently. Masters of chivalric orders had been lost before now in lesser causes.

The analogy with the statesman-patron of twentieth-century exploration is also weakened by the fact that it is hard to conceive of other princes or statesmen, let alone a modern president, being the most efficient, or even an adequately competent, commander of a vessel of discovery; while Prince Henry on the other hand was the recognized superior of his own sea captains in his knowledge of navigation, geography and cartography. He not only could have done their work, he probably would have done it better than they – certainly better than Gil Eannes' timid and superstitious forerunners in the twelve years up to 1434.

Also, while other royal patrons of discovery, such as King Philip II of Spain or Queen Elizabeth I of England, might encourage the adventures of a Cortes or a Drake, these monarchs not only had other responsibilities, but they had other interests. Even had they felt themselves free to venture abroad on missions of discovery – which certainly they did not feel – it was still the case that exploration was not their main motivation in life; for Prince Henry it was, and the *prima facie* temptation to go was that much the stronger.

Why then did Prince Henry never set sail on an expedition of discovery? Could it have been that the physical privations of such a trip, whether on a *barca* or a caravel, deterred him? This hardly seems likely when it is remembered that he habitually astounded his contemporaries by the frugality of his life at Sagres, by the desire for physical mortification that led him to wear a hair-shirt below his doublet, to fast frequently and to maintain total abstinence from wine.

Could it have been physical timidity? This can surely be discounted completely in the hero of Ceuta, who was to demonstrate on two later occasions in his life a capacity to face prolonged physical danger not only with equanimity but with such exemplary courage that it was noted by all observers.

Could it have been fear, in those around him, of the effects of his capture by the Moorish privateers during the outward or homeward voyage? This too seems an improbable cause. The *barcas* were likely to have been armed and we know that the later caravels were; there are no accounts of Henry's captains falling into Moorish hands – although the risk undoubtedly existed. But risk of capture appears to have been an acceptable risk, even for royal princes. Henry on a number of occasions was to allow himself to become surrounded by Moors in battle, and his brother Fernando was to be traded as a hostage. Capture was not an inconceivable disaster.

Could it have been that there was some social stigma attached to seafaring or exploring? Was it considered not only uncomfortable and dangerous but degrading? The contrary appears to have been the case in Portugal. The chronicler Zurara goes out of his way to describe the two squires, whom Baldaia sent out on horseback to seek prisoners, as of noble birth and people whom he knew – in one case from personal acquaintance – to be of high social standing. Most of Prince Henry's captains were young courtiers from the royal households and, both in Spain and Portugal, exploration and conquest were to become an acknowledged route to honours and titles as well as to wealth.

But perhaps the key to the problem does lie in this last direction. None of the sea captains of Portugal, Spain or England were themselves *great* noblemen, far less Princes of the Blood. Even a century later, Sir Walter Raleigh was not to be bracketed with Lord Howard of Effingham, nor Pizarro with the Duke of Medina Sidonia. Certainly in the fifteenth century the whole fabric of medieval society demanded that the truly great of this world should be seen to be so. The son of a king or the brother of a king required to keep a certain state. Prince Henry himself much relaxed the normal pattern of protocol by setting up court on a wind-swept and remote promontory, but it still was a court of a kind. Even such a Spartan court would have been impossible within the narrow confines of a *barca* or a caravel. To pass a night campaigning with his troops (and even then a silken tent would have been provided) or even to spend a few nights on shipboard en route for Ceuta, or elsewhere, was one thing; but to spend months at close quarters on terms of intimacy with common seamen would have been quite another.

Once more it was the medieval side of his nature which controlled Prince Henry's actions. 'Take but degree away, untune that string and hark what discord follows.' The natural order of medieval society made it unthinkable that he should subject himself to the indignities of a long sea voyage and Prince Henry – with his chivalric respect for that order – did not question its canons, although every fibre of his Renaissance curiosity must have cried out to see for himself the lands he longed to explore.

Chapter 6

The Tangier Débâcle

'Then the Moors by this aware
That bloody Mars recalled them there,
One by one, and two by two,
To a mighty squadron grew . . .'

Lord Byron

During the years between 1420 and 1435 when Prince Henry had been absorbed in these questions of navigation and exploration, he had spent little time at Court. A few occasions, however, had required his presence. The marriage of his eldest brother, Duarte, to Leonora of Aragon in 1428 had been one such event. (Since King João had been unable to attend the wedding – presumably through ill-health – Prince Henry wrote an account of the ceremony for his father; this was enlivened by a description of the bride fainting and being resuscitated by Prince Henry dashing cold water over her.) The following year Prince Pedro, who had finally returned from his extensive travels and campaigns in Hungary and elsewhere, married Isabel, a daughter of the Count of Urgel who was a claimant to the throne of Aragon which was occupied by the father of Leonora, Prince Duarte's wife; the marriage understandably did not improve relations between Leonora and Prince Pedro.

An even more compelling reason for Prince Henry to visit the Court during this period was the death of his father, King João, in 1433. The old King was buried in the magnificent Abbey of Batalha which he had himself erected to commemorate his victory over Castile at nearby Aljubarrota. Duarte succeeded to the throne, full of doubts about his own capacity, despite the fact that he had been effectively managing the kingdom in his father's old age for several years. But none of these events seriously or lengthily interrupted Prince Henry's concentration on the African coastline. Only one thing could do that: the call of the chivalric life, the mirage of Ceuta.

It was Henry's youngest brother, Prince Fernando, who set in motion the events which led to the Tangier expedition. Fernando had been only eleven years old when Ceuta had been captured and had never had an opportunity to win his spurs in battle. Pious and scholarly by temperament, this lack of military reputation had never unduly worried him until after the death of his father. Then King Duarte had appointed Fernando to the position which their father had held in his youth – the Mastership of the Order of the Knights of Aviz. Now he was both involved in keeping up considerable state, for which the funds of the Order were inadequate, and in setting himself up as a model of knightly conduct. The solution to both problems appeared to Fernando to lie in foreign service; abroad he would be less encumbered with the obligations of his office and more inclined to find some worthy enterprise in which to win military honours. But King Duarte was reluctant – immediately after his accession – to permit his brother to leave the kingdom, lest it should appear that there were differences between them. The King consulted Prince Henry about their youngest brother's military aspirations and Henry immediately suggested a Portuguese expedition against the Moors in Morocco, on which he – as Governor of the Order of Christ – would accompany Fernando – as Master of the Order of Aviz.

The whole debate which led up to the expedition is remarkably well documented, because King Duarte invited written opinions on the advisability of the project from all those around him. Being of a literary rather than an active disposition himself, he hoped to find in the counsel of others a buttress to his own shaky resolution. In the event, he collected so much conflicting advice that his decision was rendered all the more difficult.

The Counts of Arraiolos and of Ourém (the sons of the King's half-brother, the Count of Barcelos) both wrote in favour of the

King sending Prince Henry to help the King of Castile to oust the Moors from Granada but, while the Count of Ourém was also in favour of an expedition against Tangier preferably led by King Duarte himself, the Count of Arraiolos was against any such venture on the African continent. Their father, the Count of Barcelos, had written to the old King, a few months before the latter's death, also arguing against any expedition to Tangier.

The views of Prince Henry and Prince Pedro were already revealed. The former was actively encouraging his younger brother Fernando in his desire to take part in a military campaign in Africa, while Prince Pedro was as firmly against the project. Prince Henry was as always relentless in pursuit of his objective. He knew that Queen Leonora was antagonistic to Prince Pedro, largely because of resentment that the latter had married the daughter of the pretender to her father's throne. Henry therefore solicited her help to persuade the King in favour of the Tangier project. It was in the midst of the discussions about this project that Prince Henry decided to adopt the younger son of Queen Leonora (another Dom Fernando) as his heir. Prince Henry had already declared to the King that he had no intention of marrying; indeed he had used the argument that the money which might have been required in connection with his own marriage could be put towards a crusade against the Moors. But the timing of his decision to endow his nephew – in March 1436 – strongly suggests that the action was, at least in part, designed to consolidate the support of the Queen for his project. Dom Fernando was a favourite son of the Queen and not being the eldest would not have been in line to inherit from his father; the munificence of an uncle who enjoyed revenues from the estates of the Order of Christ, from the Algarve and elsewhere was therefore a considerable inducement. The Queen now became a leading advocate of the Tangier expedition.

Confused by conflicting advice and frightened by the responsibility of decision, King Duarte called a Council at Leiria – eighty miles north of Lisbon – in August 1436 to debate the project. Prince João, the brother between Henry and Fernando in age, spoke first. He put both sides of the case: starting by referring to the duty to crusade and the call of honour, he went on to enumerate more practical considerations. He pointed out that conversions from Islam achieved by force were of little true value, that indulgences granted for crusading could less painfully be acquired by payment, that miracles performed on crusades were often matched by miracles on more secular campaigns, that any expedition against Tangier would involve much suffering and might even put the security of metropolitan Portugal at risk. But he ended his speech by affirming that even if good sense were against the project, honour would require its fulfilment. The conclusion did not carry much conviction.

Next the Count of Barcelos, a hardened warrior now nearly sixty years of age, spoke. He found the practical objections enumerated by Prince João were the 'true flower' of reason and roundly condemned the proposal to attack Tangier.

After him, Prince Pedro rehearsed his arguments against the project. He laid particular emphasis on the impossibility of garrisoning any conquests made in Morocco, on the difficulties of financing any expedition without either weakening the currency or imposing oppressive taxes, and on the risks that the rest of Islam would come to the defence of Tangier if it were attacked by the Portuguese. It was a well-marshalled condemnation of the whole project.

Even after this weight of counsel against the expedition, King Duarte still found himself unable to resist the urgings of Prince Henry and Queen Leonora. He therefore addressed himself to the Pope, doubtless hoping thereby to shift once more the burden of responsibility for the decision from his own shoulders. Prince Henry, who knew the letter had been sent to Rome, re-exerted his influence on the Queen to such effect that when she was again in childbirth in September 1436 she extracted a firm promise from the King that a crusade against Tangier should be launched.

The Pope's reply, when it arrived after this decision had been taken, must have further discomforted the King since, far from urging on a campaign against the Infidel, it made a clear distinction between justifiable crusades to retrieve Christian lands overrun by Islam, and adventurous enterprises aimed at conquering traditionally Moslem lands. Only in the former case were Christian monarchs justified, in the Pope's view, in taxing their peoples to wage war. Tangier had never been Christian: unlike Granada, therefore, it was not an unequivocal target for Christian attack. But the decision had been taken and Prince Henry made sure there was no back-sliding.

How had Prince Henry managed to carry everything before him? His own arguments in favour of the Tangier expedition are set out in one of the very few documents of his which survive: it is a letter* written to King Duarte from Estremoz in 1436. It is by no means the most convincing of the various dissertations on the subject to be presented to the King. Prince Henry begins by listing the aims of this life as the salvation of one's own soul; the honour of one's person, name, lineage and country; the achievement of 'a joyful body', and – lastly – material gain. He then emphasizes the importance of the first two aims, particularly the achievement of honour, and goes on to denigrate the importance of such worldly delights as eating, drinking, singing, laughing and the company of women. From this he argues that the pursuit of salvation and honour being paramount, the proposed expedition against the Moors is to be pursued. He refers to the success of the Ceuta enterprise and confidently expects the Almighty to show his favour towards the newly projected expedition; he quotes numerous Biblical precedents for facing danger and argues, in effect, that 'if God is for us, who can be against us'. His only remark relating to the concrete facts of the case is a totally misleading reference to the military weakness of the Moors.

This letter can have done little to convince the King of the wisdom of the expedition; for the success of Prince Henry's cause we must look rather to the force of his personality and the influence of Queen Leonora than to the persuasiveness of his arguments. But if the letter tells us little about how Henry achieved

*Biblioteca Nacional de Lisboa, Ms I, 6–45 da Cartuxa de Evora.

his objective, it does none the less illustrate very clearly the frame of mind in which he went into the enterprise.

It was a strange reversion to his earlier self. It was twenty years since Ceuta and more than fifteen since he had attempted to lead an expedition against Gibraltar. In the intervening years he had shown himself possessed of an enquiring spirit, which had prompted his determined exploration of the Atlantic islands and the African coast. He had demonstrated an empirical approach to problems of navigation and ship construction. And he had displayed a realistic grasp of economic considerations in the colonization of Madeira and the management of his affairs. All this suggested qualities which we associate more with a Renaissance than with a medieval personality. Yet now he deserted his projects at Sagres, propounded arguments to his brother that had the ring of a friar preaching a medieval crusade, and set out on a course of devious, obdurate and reckless behaviour which was to have disastrous consequences for all associated with him.

From the beginning, the preparations for the expedition did not go well. It was calculated that an army of 14,000 would be required to take Tangier, but despite energetic recruiting efforts, which included granting qualified pardons to a somewhat arbitrary range of criminals (woman-garrotters, for instance, were protected, but only in the district of embarkation), the numbers fell far short of the target. Transport, too, was a problem and attempts to charter ships from England, the German cities and Flanders all met with failure. Nor did the Portuguese manage to keep the destination of the expedition secret, as they had done in the case of Ceuta; Tangier strengthened its defences and kept in close touch with the neighbouring Moorish kingdoms of the Maghreb.

Finally, however, all was as ready as it ever appeared likely to be. A force of 2,000 cavalry, 1,000 crossbowmen, and 3,000 infantry and archers – a mere 6,000 in all – was assembled. A ceremony took place at the Carmo Church in Lisbon, after which a procession, led by the faded banner of the former Constable Nun' Álvares Pereira, wound its way to the shore. Everything was done to emphasize the parallels with the Ceuta expedition. The fleet moved downstream to Belem and eventually sailed on the 23rd of August 1437.

Three days later they arrived at Ceuta, which was still under the governorship of the same Count of Viana who had been the host and companion of Prince Henry when he had stayed there in 1418. In the intervening years the Governor had repulsed numerous attacks and endured almost continuous harassment from the Moors. Frequently he had had to take the field twice in a day and he was reputed never to have shed his coat of mail for sixteen years, 'so that it split in several places, as if it were of cloth'. Foreigners from all over Europe, including an uncle of the Holy Roman Emperor, had come to join his garrison for long enough to see some service against the Infidel and establish their military credentials.

No one could accuse the Count of Viana of being easily daunted by greater numbers or reluctant to fight. Yet he and his Council were dismayed at the smallness of the Portuguese army and urged Prince Henry to stay at Ceuta, either until reinforcements could be assembled or until they could agree to modify their objectives. Neither scheme was acceptable to Henry, who urged that the fewer the crusaders, the greater the share of the glory. He recalled similarly alarmist advice which later had proved to be unfounded before the attack on Ceuta. He overbore the Council and set out to march via Tetuan to Tangier.

King Duarte, with his customary propensity for written advice and instructions, had prepared in his own hand two directives for Prince Henry; one dealt with moral questions and the other with specific operational orders. The King had urged Prince Henry to read the latter often and heedfully; from the outset Henry disregarded this command. The King had directed that the fleet, which was to remain with the Princes and supply their needs, should be divided into three squadrons which were to go to different points on the coast, thus attempting to regain some element of surprise about the destination of the Princes' army. Prince Henry sent the whole fleet direct to Tangier. Prince

Fernando remained on board rather than march with the army as he was suffering from fever.

When on the 13th of September Prince Henry and the army reached Old Tangier, the Roman ruin just outside the Moorish city, they found it deserted. Still obsessed with the ease of the conquest of Ceuta, they speculated that perhaps the Moors had abandoned the city itself. An immediate assault was launched and bloodily repulsed. Prince Henry's banner when unfurled before the walls of Tangier was seized by a gust of wind and torn to shreds – an ill omen, his men predicted, especially on Friday the 13th.

Prince Fernando had rejoined the main army in front of Tangier and together the Princes established a camp and spent two weeks in unloading the siege equipment from the fleet. The Moorish garrison, it transpired, was commanded by the same Sala-ben-Sala who had commanded at Ceuta when the city fell: surely the Princes argued, he would be doomed again. A futher assault was made on the 27th of September, but although Prince Fernando, the Count of Arraiolos (who had not allowed his initial disapproval to keep him from participating in the expedition), the Bishop of Ceuta and many others displayed exemplary courage, it was found that the scaling ladders were too short and the defences too good. The attack was repulsed with 500 of the assailants wounded and twenty dead. Prince Henry sent to Ceuta for longer ladders and better siege equipment.

But meanwhile a much more sinister development than the repulse of this assault was taking place. In the weeks that had passed since Sala-ben-Sala realized that an attack was imminent he had sent word for help to the King of Fez and the King of Beles; he had even sent as far as the King of Morocco (Marrakesh) who commanded the allegiance of the Berber tribes inhabiting the fringes of the Sahara. While Prince Henry and his army were sending for better equipment and skirmishing with the enemy in preparation for a fresh assault, these neighbouring kingdoms were despatching their warriors to the aid of their fellow Moslems besieged at Tangier.

When longer ladders and a siege-tower had arrived from Ceuta, and when yet more ladders had been made locally, Prince Henry prepared for a third attack. But he found that there was insufficient powder to fire the fifty cannon balls of stone, so further delay ensued while he sent another ship to Ceuta. When eventually all was ready a fresh assault on the walls was made. However, the Moors managed to set fire to the wooden siege-tower, to burn or break the ladders and to scald the assailants with the traditional boiling oil.

There was another and possibly more decisive reason for the failure of this attack. Prince Henry now had to divide his meagre forces and leave nearly half behind to guard his own camp. In fact, the hills around were by now dark with the menacing hordes of Arab and Berber tribesmen who had responded to the call for help from Sala-ben-Sala. While Prince Henry attacked the city the tribesmen attacked his camp. Both attacks were repulsed.

Prince Henry had now made three full-scale assaults on the walls of Tangier. The written directives from King Duarte explicitly commanded him to withdraw to his ship without delay, and to proceed to Ceuta where he should await further instructions or reinforcements, after three unsuccessful attempts. The King was well aware of the danger of the Portuguese force being swamped by Moslem reinforcements. Both the order and the reason for it must have been patently clear to Prince Henry, yet he chose to disregard it.

In the following few days the situation deteriorated rapidly. During particularly heavy fighting on the 9th of October, Prince Henry – wearing as always his heavy black suit of chain mail – had his horse killed under him and was only saved by a page who managed to get a fresh mount to him. The Portuguese were now on the defensive. The spears and lances of the Arabs and Berbers glinted on the hills all around them. Contemporary estimates, which were almost certainly exaggerated by chroniclers anxious to justify subsequent events, put the total strength of the forces now mustered under the green banner of Mohammed at 700,000. A more credible earlier estimate had referred to

40,000 horsemen. At all events, the badly battered remnant of Prince Henry's 6,000 was in desperate straits.

It was now that Prince Henry disobeyed the third of the injunctions put upon him in King Duarte's written directive. On no account, he had been told, should he allow his force to lose contact with the ships that brought them. Yet Prince Henry had sited his camp too far from the coast, in fact about a mile away it seems. Now the Moors surrounded his camp and severed his links with the ships. This was a dual disaster: first, no fresh supplies could be received; and second, the only possible line of retreat was cut. Even if Prince Henry had not contemplated the possibility of retreat (and it was rash indeed if he had not), he who had so assiduously supervised the provisioning of other expeditions should surely have made certain that he had adequate supplies of food and water in his own camp. But this too appears to have been overlooked. Within a few days the Portuguese troops were reduced to sucking fresh water from the sand after a providential shower, and slaughtering their horses for food. They roasted the horse-flesh on fires kindled from the padding of their saddles. The heat of the sun on the bare plain was intense. The attacks of the enemy were continuous, those of the surrounding tribesmen being supplemented with savage sorties from the walled city of Tangier.

In these circumstances Prince Henry's stubbornness, which had contributed so largely to the predicament in which the Portuguese army found itself, proved a strength and inspiration to those around him. All the subsequent reports agree that he did not spare himself: encouraging the infantry and bowmen at the hastily improvised barricades, hazarding himself in the thickest parts of the fray, he was as gallant a commander as he had been reckless a general.

On the 10th of October the decision was belatedly taken to try to cut their way out of the encircling Moors by a night march to the sea. But one of the chaplains to the Portuguese army preferred to risk his Christian soul than endure the privations and dangers of remaining with Prince Henry, and he betrayed the Portuguese intentions to their enemies. In consequence the Moors were on the alert to withstand the attempted break through their cordon, and the Portuguese were repulsed and obliged to remain in the trap.

The only alternative to total annihilation now lay in negotiating surrender terms. The bitterness of such a conclusion for the Prince, who had been brought up in the tradition of the Templars, who carried their cross on his banner, who had triumphed at Ceuta and who had embarked on this crusade in obedience to the chivalric strain in his nature . . . such bitterness can well be imagined. And the terms themselves were harsh indeed. The Moors were prepared to allow the Portuguese army to regain their ships in safety, leaving their arms behind them, only on condition that Ceuta was surrendered to its original Arab masters. Ceuta was the symbol of Portuguese crusading zeal; it was their sole foothold on the African continent; it had always been regarded as the gateway to future expansion. Its surrender would close the door definitively on a chapter that had scarcely begun.

Unpalatable as these terms were, there seemed no option but to accept them. There was a further bitter pill to be swallowed. Sala-ben-Sala insisted on one of the two Portuguese Princes being handed over as a guarantor of the fulfilment of the terms; he was prepared to exchange one of his own sons as a pledge of his own faith. Prince Henry offered to go as the Portuguese hostage. But the Council would not hear of it: he was the commander-in-chief and his loss would not only be the greater humiliation for this reason, but would leave the Portuguese army without its nominated leader. The Council's advice had often – all too often – been set aside; but on this occasion Prince Henry does not appear to have resisted it long. Prince Fernando, also a willing victim, was offered up.

Prince Henry's motives require some analysis. If the exchange of hostages was genuinely expected to be a temporary and short-lived arrangement, then it clearly was more sensible to part with Fernando. Prince Henry was responsible to the King for the safety of his army; he would obviously be prevented from concluding these responsibilities if he were a prisoner in the enemy camp. But if, on the other hand, Prince Henry were already determined that Ceuta should never be surrendered, then the sacrifice of Fernando was unlikely to be short-lived. The morality of allowing his brother to be given up, as a guarantor of a course of action which he – Henry – had every intention of frustrating, would be dubious indeed. If the life of one of the two Princes was to be the price of Ceuta, then was Henry justified in choosing that it should be the life of his brother? It seems almost certain that he did not see it in these terms. To start with, he may genuinely have intended that, if the Moors kept their part of the bargain, then the Portuguese would keep theirs. And even when it became clear – for reasons that will appear shortly – that the Portuguese had a convincing pretext for considering as invalidated their promise to give up Ceuta, both Prince Henry and King Duarte probably calculated that they could negotiate the return of their brother for some lesser price, or else that they could recapture him. Their subsequent actions reveal such thinking. So Prince Henry should probably be exonerated from any charge of cowardice, cynicism or selfishness in surrendering his brother Fernando.

The exchange was carried out on the 16th of October, a number of Prince Fernando's household – some half-dozen in all – choosing to go with him into captivity. Prince Henry had never been a jocular man; from this day onwards, until the end of his life, he smiled but seldom.

The following day Prince Henry collected the army around him for the retreat to the shore and the ships. But even before they had abandoned their camp and weapons, fighting broke out again. It was not the army of Sala-ben-Sala which attacked the Portuguese but the ill-disciplined hordes of Berbers and others from the outlying parts of Morocco. To them, the prospect of a massacre and resultant plunder was irresistible and they did not feel themselves bound by the agreement between the main protagonists. Appeals to Sala-ben-Sala could not bring about a truce, since he appeared incapable of controlling his wilder allies, and eventually it was an armed force, fighting every inch of the way, which Prince Henry led to the sea. The embarkation was carried out under a hail of spears.

The main fleet, under the command of the Count of Arraiolos, returned forthwith to Portugal. Henry himself sailed for Ceuta, from whence he sent a message to Sala-ben-Sala pointing out that, as the agreed safe-conduct had not been granted to his army, the surrender terms were nullified. Ceuta would stay in Portuguese hands. He awaited the return of Fernando and would deliver back the son of Sala-ben-Sala.

A fleet was despatched to carry this message and his son back to Sala-ben-Sala. The latter, being Sheik of Arzila as well as Sheik of Tangier, had meanwhile removed his prisoners for greater safety to the former town. Prince Fernando and his companions had been roughly handled, being mounted on broken saddles (the chroniclers make much of this point, presumably being struck by the contrast with the splendour of a princely equipage), left hungry and thirsty, subjected to a forced march and reviled everywhere by the Arab population.

The Portuguese rescue party was able neither to negotiate an exchange of prisoners nor to retrieve Fernando by force. Sala-ben-Sala, it later transpired, was not particularly interested in the return of his own son. He let it be known that he had plenty of other sons and indeed had recently had one of them executed: paternal sentiment was not a factor on his side. The desire for Ceuta was. The Portuguese rescue fleet returned dejected to the Algarve while Prince Henry lingered at Ceuta.

Just as a great debate had taken place to assist King Duarte in reaching his decision about whether or not to attack Tangier, so now another great debate ensued to help the indecisive King over the question of exchanging Prince Fernando for Ceuta. As was his custom, he asked advice from one and all

Prince Pedro, who had consistently maintained that possessions in Africa were more of a liability than an asset to Portugal, urged surrendering Ceuta. In this advice his affection for his brother Fernando, and indeed his affection for the whole concept of a united brotherhood of Princes, must have been a potent factor. Prince João concurred with his brother's advice.

There was no shortage of contrary voices. The Count of Arraiolos, having fought bravely at Tangier, was now outspoken on military grounds against any surrender of conquests already made. The Archbishop of Braga was also against surrendering Ceuta, on ecclesiastical grounds; he pointed out that the return of the city to the Moors would mean the desecration of Christian churches. The burghers of Lisbon and Oporto were also against a surrender, on the predictable grounds that Ceuta – even if not a very valuable trading post at present – had a potential commercial value.

The Cortes was convened at Leiria and discussed the question extensively and inconclusively. It was suggested that the circle of advice might be widened to include other Christian monarchs. All those approached replied in sympathetic terms, applauding the self-sacrifice of Fernando; none offered any material assistance for his release. An attempt to involve the Moorish King of Granada as an intermediary met with no greater success. Next, as in the initial debate about the expedition, the Pope was invited to give his views. He refrained from reminding King Duarte that he had had grave qualifications about the justification for the adventure in the first place, but he too gave no promise of assistance to rescue Fernando, and suggested that the surrender of Ceuta would indicate a lack of respect and honour to God's realm on earth.

Finally the King sent for Prince Henry. The latter had delayed for five months in Ceuta and had then returned to his lonely retreat at Sagres. The Court was at Evora. Slowly Prince Henry rode north, but when he reached Portel – some eighteen miles south of Evora – he sent word to the King asking to be excused from attendance at Court. The King rode out to meet his brother at Portel in June 1438: the commander-in-chief had not rendered account to his sovereign for eight months after the military disaster. Now that the moment had come, Prince Henry was unrepentant. He talked about raising a far larger army – of 24,000 men – and launching another campaign against the Moors in Africa; he viewed the surrender of Ceuta as unthinkable, and expressed regret that it was not he who was the hostage. He returned to Sagres.

King Duarte was overwhelmed by the dilemma. All his own instincts were in favour of following Prince Pedro's advice and shedding the complication of Ceuta to be reunited with his dearly beloved youngest brother. Yet such was the force of Prince Henry's personality and determination that the King could not bring himself to override his counsel. There were outbreaks of the plague at Evora and the Court dispersed, the King going to Tomar. By August he had physically collapsed: some attributed his fever to plague and others to his worry over the fate of Fernando, who had written to the King of his sufferings. Whatever the cause, on the 9th of September 1438 he died, at the age of forty-seven.

What the honourable but vacillating King had been unable to decide in his lifetime, he thought he had decided by his will. This charged that his private fortune should be devoted to ransoming Fernando, and if the sum proved insufficient it was his wish that Ceuta should be handed over. Prince Fernando did not know the terms of the will, but he was informed of the death of his eldest brother and drew the conclusion that the prospects for his release had been greatly diminished by the decease of a brother and King whom he knew to be devoted to him.

Indeed, his own condition had greatly worsened since his initial captivity, and was to deteriorate far more. The King of Fez had been afraid that Sala-ben-Sala might trade the return of Fernando for some lesser price than Ceuta. He had therefore requested the transfer of the captive to Fez. This had been greatly dreaded by Fernando since the vizier of the Kingdom of Fez – Lazeraque – was notorious for his cruelty and fanatical detestation of Christians. The journey from Arzila to Fez had been undertaken in May 1438 and had been the occasion for six days of exposure to insult and threats of violence on the road between the two towns.

Once arrived in Fez, Fernando was placed in the charge of the dreaded vizier. After some months of confinement, during which a foreign merchant managed to smuggle food to him, he was allowed once more into the fresh air. But the benefits were limited: he was forced to labour in the vizier's gardens and groom his horses, while laden with chains. He shared a cell, built for eight, with eleven other prisoners. Sheep skins and hay were their bedding. Food was restricted to two loaves of bread a day. But his Portuguese companions were with him and life was endurable; besides, although he did not know it, a move was afoot to effect his release.

What King Duarte could not bring himself to do, Prince Pedro and King Duarte's widow (who were now joint Regents) achieved: they sent an envoy to Arzila to negotiate the exchange of Prince Fernando for Ceuta. It is not clear quite what prevented the achievement of this object. One reason was undoubtedly the vizier Lazeraque's reluctance to part with his captive; he appears to have hoped for a monetary ransom to be paid to himself, and delayed sending Fernando back to Arzila. Another reason was that the Moors declared they would not part with Fernando until Ceuta was in their hands, and the Portuguese did not trust them sufficiently to hand over the city in advance of receiving their hostage. But it seems likely that there may have been a further reason: Prince Henry was still averse to the deal and his influence may well have dampened the enthusiasm and vigour with which the Portuguese emissaries pursued their task.

Whatever the reasons, Fernando remained a captive at Fez. Now his durance became vile indeed. The Prince was no longer allowed to go out to work with his companions, and when it was noted that he still derived some pleasure and comfort from their company after their return from the fields, he was moved to a separate dungeon. This was a punishment cell so small that he could not stand up; it was riddled with vermin; it stank from an adjoining latrine used by the eunuchs of the vizier's household.

The fortitude with which Prince Fernando withstood these ordeals was recorded by his chaplain and confessor – Friar João Álvares – who was subsequently ransomed and wrote an account of the imprisonment of his patron. It is natural that such a record should emphasize the Christian piety of its subject, but in the case of Fernando this quality was recognized even by his gaolers. He did not complain of the malicious rigours imposed on him and even offered up prayers for the misguided souls of his captors.

It appears that dysentery was the final cause of his death on the 5th of June 1443. When the squalid conditions of his confinement are considered, it is remarkable that this apparently sensitive and previously cloistered Prince, who had been ill even during the initial campaign, should have survived so long. He had spent fifteen months in the punishment cell and nearly five years as a captive of the Moors. Even when he was dying little consideration was afforded to him beyond the attendance of his own physician and his confessor. After death, no consideration at all was to be shown: his body was hung by the feet from the city walls of Fez.

His Portuguese companions, most of whom were also to die in prison at Fez, managed to remove his heart and embalm it. Those who were subsequently ransomed, eight years later, brought the sacred relic to Portugal to be buried in the tomb allotted to Prince Fernando at Batalha Abbey (where it was joined twenty-two years later by the remainder of his corpse which had by then been retrieved from Morocco). As the procession with the 'Martyred Prince's' heart passed on its way to Batalha, Prince Henry joined it. He attended the solemn martyrs' service and paid his last respects to his youngest brother.

He had indeed much to answer for. It was he who had urged a foolhardy expedition. It was he who, disregarding both his orders and good sense, had brought about the destruction and humiliation of the Portuguese army. It was he who had acquiesced in the surrender of his brother Fernando and subsequently had hardened his brother Duarte's heart against negotiating Fernando's release. Both Duarte and Fernando were now dead. But disaster and tragedy had not softened the forcefulness of Prince Henry's character; he was still to pursue and persist in those objectives which awakened his remarkable imagination; and those objectives were still to lurch between the twin concepts of medieval chivalry and Renaissance enlightenment that constituted his own peculiar nature.

Tangier a century after the Portuguese attack

Lagos harbour, from which Prince Henry's caravels sailed and where the first slave market took place

The Mariners' Card at Sagres

Chapter 7

On the Rocks of Sagres

'Though Truth and Falsehood be
Near twins, yet Truth a little elder is.
Be busy to seek her; believe me this:
He's not of none, nor worst, that seeks the best.'

John Donne

Prince Henry's interest in Sagres and inclination to spend his time there dates from 1419 when he returned from his second trip to Ceuta and when the King appointed him Governor in perpetuity of the Algarve. But it was only after the humiliation and disillusionment of the Tangier expedition, in 1437, Zurara tells us, that 'he commonly remained there' and, although there is disagreement among scholars as to the date of commencing the construction of the so-called *Vila do Infante* at Sagres, it seems likely that work began about that time. Lagos was still the main port for Prince Henry's vessels of discovery; it was there that they recruited their crews and it was there that they subsequently sold the merchandise they brought back. But Sagres itself increasingly became the scene of Prince Henry's own activities, of the planning of the trips, of the studying of charts and instruments, of the consulting of visiting savants or resident experts.

What sort of place was Sagres and what precisely went on there? The first part of the question is more easily answered than the second. The Sagres peninsula juts out into the Atlantic on the extreme south-west corner of Portugal and of Europe – being two or three miles less far west but farther south than its twin promontory of Cape St Vincent. The country between Sagres and the seaport of Lagos, twenty miles farther east along the southern shore of the Algarve, is bleak indeed. The Atlantic winds sweep the rocky, flat, stony land and – as has been remarked in an earlier chapter – Prince Henry could hardly have found a more inhospitable corner of the country in which to pursue his activities. But it was geographically the closest point in Europe to those sea lanes which he was opening up to the West African coast; he could, in the words of the German philosopher Mickle, in 1775, 'contemplate the ocean from his window on the rock of Sagres.'

But apart from contemplation, what was the nature of the activity at Sagres? Earlier biographers of Prince Henry wrote of his School of Navigation at Sagres and suggested a more formal academy of sciences than seems likely, in the light of more recent scholarship, to have existed. Possibly such speculation reflected the seventeenth- and eighteenth-century idea that all learning was best organized around such bodies; the age of the foundation of the Royal Society left a social imprint on scientific thought that inclined to see institutions wherever it saw progress. The nineteenth-century historians saw Prince Henry as a scientist and an advanced student of mathematics, navigation and cartography; they extrapolated from their own ideas and imposed them on the slender historical evidence. Contemporary documents provide no confirmation. It remains the case, however, that substantial and constructive studies were undertaken in the fields of ship-design, navigation and cartography, and in directing these activities Prince Henry demonstrated clearly and decisively those characteristics which we have come to associate with an empirical Renaissance thinker. How he did this can best be seen by reviewing each field of activity in turn.

In the field of shipbuilding there was a clear need to solve a specific problem. One of the main impediments to the exploration of the West African coast was the difficulty of the return voyage. The further down the coast the Portuguese ships sailed, the greater the problems and delays attendant on their return, because the north-westerly winds persisted and thus the whole of the return voyage was spent sailing into the wind. With provisions running out, the hazards of this slow return by *barca* or *barinel* (the only two types of vessel – as we have seen – which were available to Prince Henry's captains at this time) were a major and adverse factor in the whole exploratory process. Neither ship could sail closer to the wind than an angle of 67 degrees. It was true that the *barinel* had oarsmen to help propel the ship, but in the heat of the Saharan coast this, too, was an inefficient method of propulsion.

Prince Henry had pondered on this problem, had listened to reports reaching him from the eastern Mediterranean and had observed the shipping in his own native city of Oporto. The Arabs off the Egyptian coasts since antiquity had used small dhows with lateen (i.e. triangular and slanting) sails; off the Tunisian coast such boats were known as *caravos*, a name apparently originating from the tiny Greek fishing vessels made of rushes and hide, and were generally about forty-five foot long by nine foot in the beam. Some of these vessels, however, were reputed to carry loads of up to seventy horses and crews of thirty Arabs, and must therefore have been considerably larger. On the Douro River in the north of Portugal a similar style of vessel was also in use in Prince Henry's time. Much smaller than the Arab *caravo*, this Portuguese vessel was called by the same name with the addition of the vernacular diminutive *ela*: it was a *caravela*. Reference is made to these craft by that name as early as 1255 in a charter granted to Vila Nova de Gaia, the riverside quarter of Oporto (now the centre of the port wine shipping industry).

Prince Henry put these two pieces of knowledge together. He persuaded his shipbuilders to design a craft with the manoeuvrability of the Douro River *caravelas* and some of the freight-carrying capacity of the Tunisian *caravos*, and in doing so they produced a compact and agile vessel which was to be the mainstay of the Portuguese explorers for the next century. An axled rudder, borrowed in design from the shipbuilders of northern Europe, completed the composite design. The earlier models had two or three masts, all with lateen sails, and with a displacement of about fifty tons. They were generally fully decked, with little or no raised superstructure forward but a raised stern comprising a poop cabin, probably occupied by the captain and any knights who were on board. The rest of the crew, who never numbered more than about twenty, would have slept on deck, or huddled below among the cargo in bad weather. The later models, which were to be developed after Prince Henry's lifetime, were larger and frequently had a fourth mast forward (which was square-rigged) in addition to the three lateen-rigged masts; they were known as *caravelas redondas* and frequently displaced up to one hundred and fifty tons.

The shape and angle of the lateen sails enabled Prince Henry's caravels to sail a course much closer to the wind. Instead of 67 degrees from into the wind, they could sail at 55 degrees (see diagram). Over a voyage of many hundreds of miles up the West African coast, the twelve degree difference could mean a saving of many weeks and an enormous increase in the confidence of the captain and his crew in their capacity to make good their return. From 1441 onwards, the rate of progress down the coast greatly accelerated, in no small measure due to this enhanced confidence.

The design of the caravel was broad in the beam; as distinct from its narrower Tunisian antecedent, the caravel of the 1440s tended to be as much as twenty foot in the beam with a length of about sixty foot. This did not, in fact, produce a great improvement in comfort for the crew, but it did allow of a shallow draft which was to prove invaluable when exploring the unknown inshore waters of the African coast. It also made it possible, as will be seen in a later chapter, to beach the vessel – for careening or repairs – with a minimum of difficulty.

Prince Henry's practical nature did not allow him to limit his activities to the theoretical. Lagos was developed into a shipbuilding centre to rival Lisbon and Oporto. The pine wood which was used for the hulls of the caravels was grown in profusion along the Atlantic seaboard of Portugal, where pines were protected by law in the fifteenth century both on account of their value as a corrective to the erosion of the sandy soil and as a valuable strategic commodity. The resin of the cluster pines was

Progress sailing into the wind
by a square-rigged Barca

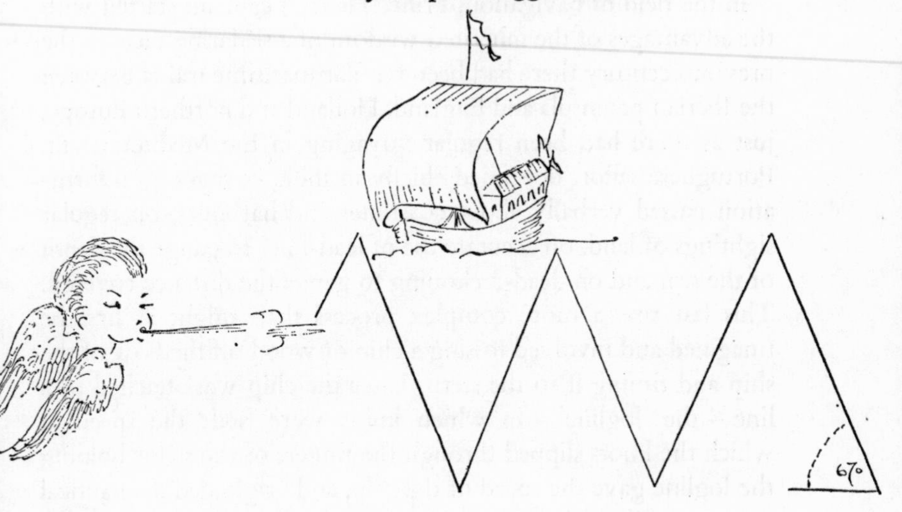

by a lateen-rigged Caravel

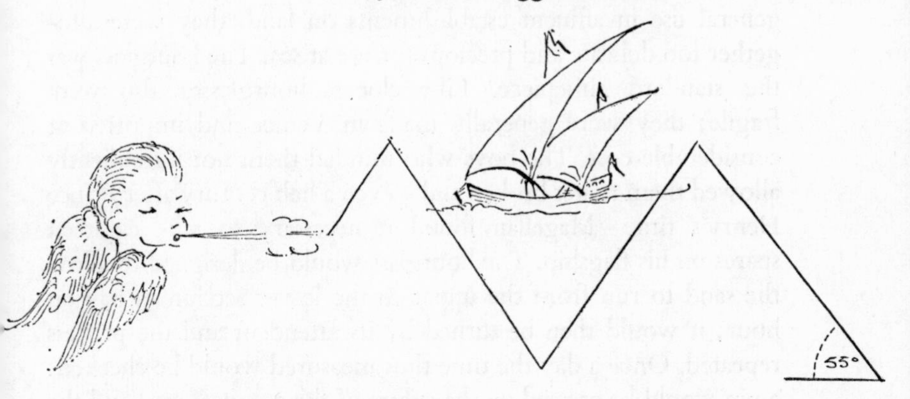

also utilized to caulk the hulls. Oak grew in plenty in the forests of the Alentejo and was used for the keels. Sailmaking and rope-making thrived in the proximity of the shipyards of Lagos.

In the field of navigation, Prince Henry's captains started with the advantages of the inherited wisdom of a seafaring race. In the previous century there had been regular maritime traffic between the Iberian peninsula and England, Holland and northern Europe, just as there had been regular voyaging in the Mediterranean. Portuguese sailors depended chiefly in these voyages on information passed verbally about coastlines and harbours, on regular sightings of land, on accurate use of lead-lines to gauge the depth of the sea, and on dead-reckoning to gauge the distance covered. This last was a more complex process than might at first be imagined and involved tossing a chip of wood off the bow of the ship and timing it to the stern. Later the chip was attached to a line – the 'logline' – in which knots were tied; the speed at which the knots slipped through the fingers of the sailor holding the logline gave the speed of the ship, and originated the nautical measure of 'knots'.

To make effective use of the logline, it was also necessary to have accurate estimation of time. Although clocks were in fairly general use in affluent establishments on land, they were altogether too delicate and precious for use at sea. The hourglass was the standard timepiece. Like clocks, hourglasses also were fragile; they were generally made in Venice and imported at considerable cost. The boys who minded them not infrequently allowed them to get broken and – even a half century after Prince Henry's time – Magellan found it necessary to take eighteen spares on his flagship. The hourglass would be designed to allow the sand to run from the upper to the lower section in half an hour; it would then be turned by its attendant and the process repeated. Once a day the time thus measured would be checked: a pin would be erected on the centre of the compass card and the exact moment when its shadow reached the fleur-de-lis that marked north (Prince Henry's navigators never reached the southern hemisphere) would be noted and the glass restarted from that moment. Forty-eight hourglasses later, the process would be repeated and any grave discrepancy attributed either to having sailed a substantial distance east or west, or to rough weather having impeded the passage of the sand, or to the glass having been warmed (thus expanding the thin glass and allowing the sand to trickle through more quickly). In the last instance, the boy in charge of the glass would be suspected of having deliberately heated it to shorten his watch; dire penalties followed. It seems likely that minding the hour-glass, like cooking for their masters, would have been one of the tasks allocated to the young pages whom we know were taken on the expeditions.

The compass was also in regular use, though still held in some superstitious distrust by common seamen. In fact, the origin of the compass is made the more difficult to determine because many of its earliest European utilizers concealed their instruments for fear of being accused of diabolical practices. There was probably also a natural reluctance to share such an advantageous aid to navigation with rival seafarers. Captain Hitchins and Commander May* have pointed out that there were probably five separate stages in the discovery of the compass. First would have come the discovery of the powers of the lodestone (found in parts of the world ranging from Norway to India) to attract iron; then would have come the realization that this power could be transferred – by rubbing – from the lodestone to the iron itself; then it would have been observed that iron thus magnetized would point to the north if floated in some way; next might have come the development of placing the magnetized needle on a pivot; and finally there emerged the combination of floating on a pivot thus reducing friction and oscillation. More difficult to determine is how and where these stages of discovery were accomplished.

Claims have been made by many nations to have discovered the magnetic compass, the most ambitious being a Chinese one to the effect that in the third century BC their chariots were guided across the desert by small figures, fixed on the front of

*In *From Lodestone to Gyro-Compass.*

the chariot, with outstretched arms pointing permanently either north or south. But it was not until the twelfth century AD that references to the compass appear in European documents. Almost certainly the knowledge of magnetic needles and their use as compasses reached Europe from Arabia. An English monk, named Alexander Neckham, describes a pivoted compass in a treatise written in 1187; it is thought that he gained his information from the University of Paris where he had studied a few years before. At all events, there were frequent references to the compass in English and French literature during the following century; Roger Bacon wrote about the phenomenon and a French friend of his wrote a treatise on the subject in 1269. Progress was made in developing the compass at Amalfi in the early years of the fourteenth century and this formed the basis of some Italian claims to have invented it; it seems more likely that the main Amalfi contribution was the attachment of a compass-card to the needle. By 1345 there is a record of 'adamants or sailstones' being purchased for use on an English ship and by the end of the fourteenth century 'seyling nedels' appear on the inventories of a number of English ships.

There was therefore no doubt that the compass was well known in nautical circles by the time that Prince Henry sent out his first expeditions. But it was far from being established as a respectable instrument. Its counterpart, the lodestone, was credited with many strange qualities: it was said that if placed under the pillow of an unfaithful wife it would make her confess her sins. Other strange powers were attributed to it: the healing of certain diseases, and even efficacy as a contraceptive. The fears that had earlier led sea captains to conceal their reliance on the compass, lest they should be accused of trafficking in the occult, persisted. Even fifteenth-century navigators were sometimes required to rebut charges of devilry by the strangely medieval argument that, since the needle could best be used when floating attached to a straw which it pierced at right angles, it formed the shape of the Holy Cross and could not therefore be an instrument of Satan! Prince Henry's contribution at Sagres to: he use of th ecompass was not a scientific one; it was the establishment of an empirical and practical atmosphere where such obfuscating nonsense had no place.

With expeditions sailing ever further down the African coast, the accurate estimation of latitude became ever more urgent. Dead-reckoning, land sightings, compass and hourglass might be enough for the confined waters of the Mediterranean, but were insufficient for venturing into the unknown south. For this the astrolabe was necessary. It, too, had a long history and was known to the Arabs before it was introduced to Europe; fine examples of astronomers' astrolabes made in Baghdad around AD 830 survive to this day. But like the clock, the astrolabe was too delicate an instrument to be carried and used on one of the small *barcas* or caravels of the time. Yet some device was necessary to measure the angle of the sun from the horizon and thus estimate the number of degrees south which the mariner had reached. A simplified version of the astrolabe, known as the *balestilha*, was devised in the form of a cross-staff; this was a short staff fitted with a moving cross-piece at right angles. The mariner would point the staff between the sun and the horizon in such a way that the cross-piece aligned with the sun at the top and the horizon at the bottom. The distance between the cross-piece and the end of the staff was measured in degrees from which could be calculated the latitude. Prince Henry encouraged his captains to take this simplified instrument with them and to take bearings whenever possible from dry land rather than the rolling deck of a ship.

He encouraged the study of the night sky (which had been the subject of a passage in King Duarte's *Leal Conselheiro*) as a means of determining time, and therefore position, by the movement of the stars. During his lifetime many of the initial observations were carried out which were to result in the *Regimento de Evora*, a codification of such researches published in 1519. All such study of the heavens was directed towards practical results and away from the mystical and astrological aspects of the stars which dominated the work of most of his medieval contemporaries.

Cartography was probably a greater preoccupation at Sagres

than either shipbuilding or navigational instruments. In the fourteenth and fifteenth centuries, the study of the world's surface had much in common with the study of the heavens: both were, at one and the same time, medieval metaphysical exercises and 'modern' scientific disciplines. Just as there were astronomers whose main motivation was astrology, and who tried to interpret the stars in terms of their own philosophy, so there were cartographers whose main purpose was to rationalize the world around their own preconceived religious and philosophical views.

All maps owed much to the influence of Ptolemy (c. AD 130) and the Arab geographers who had followed him. Ptolemy himself had speculated that the world was round but had made no attempt to link a far eastern land mass, on which he had placed purely fictional features, with the Atlantic ocean which he had cut short not far from the Iberian coast. He had not envisaged any western hemisphere. Ptolemy had also linked Africa with a southern land mass and had thus precluded the possibility of travelling to the East by sea round southern Africa. His influence was enormous throughout the years of Arab ascendency, when Islam appeared to have taken over from the ancient world the custodianship of all geographical knowledge. But in the tenth century, Massoudy had modified Ptolemy's world map in some important respects, not least by postulating a channel between southern Africa and the nebulous land mass around the southern extremities of the world. His work was continued in the twelfth century by Edrisi, another Arab geographer, who enjoyed the patronage of a Christian monarch – King Roger of the Norman Kingdom of Sicily. Edrisi recorded the travels of his contemporaries throughout the Christian and Moslem worlds, even including an account of the adventures of the *Maghrurins* – some 'wanderers from Lisbon' who reputedly lighted on an Atlantic island during the early part of the twelfth century.

From this common basis of classical and Arabian knowledge, the twin tendencies of medieval European geographers and cartographers sprang: the tendency to construct a symmetrical design of God's world in harmony with metaphysical beliefs, and the contrary tendency to construct a realistic chart of positive and proven discovery and knowledge. Prince Henry would have been familiar with both tendencies. In the former category he would, for instance, have been familiar with the Psalter Map of the thirteenth century; this illustrated the world with Christ at the top and dragons crushed beneath him at the bottom, with Jerusalem as the navel of the earth in the centre, and with various fanciful legends inscribed in place of geographical data. Doubtless, Prince Henry would also have been familiar with the more famous Hereford Map, completed at the beginning of the fourteenth century, which also placed Jerusalem as the centre of the world and filled large spaces with representations of monstrous animals (including unicorns); this map tried to find a place for most bizarre travellers' tales – based on the works of Pliny and others – rather than for the findings of reputable mariners.

In the category of more scientific cartography, he would have been familiar with the Venetian map of Marino Sanuto which attempted in 1306 to put into cartographical form Italian ideas regarding the Atlantic coastline of Africa, ideas which had prompted the Vivaldi brothers in the previous century to make an expedition down the West African coast from which they had never returned, thus adding force to the theories that the sea beyond Cape Bojador was not navigable. This Venetian map shows a sea route to the south of Africa, but a conjectured rather than a discovered one.

Another and somewhat more 'scientific' map with which Prince Henry would have been familiar was a Florentine one of 1351 known as the Laurentian Portolano. The Portolanos were coastal charts and this one incorporated records of discoveries made a few years before about the north-west African coast and its islands; it then went on to extrapolate a coastline further south and guess – with a surprising degree of accuracy – how the coast would later be discovered to be indented at the Gulf of Guinea and then to carry on round the southern cape of Africa into the Indian Ocean.

He would also have known of other Italian maps, among them

the Venetian ones produced by the Pizzigani brothers in 1367, that produced by the Camaldolese convent in Murano in 1380, and the chart made by Andrea Bianco in 1436. All these emanated from the practical school of geographers. But more valuable to him than any of these was probably the Catalan map of 1375 which reveals a remarkably accurate knowledge of the western Sudan (Sahara) region, including the caravan routes to the 'land of the Negroes', and which also incorporated the findings of Catalan explorers who had undertaken in 1346 an unsuccessful voyage in search of the 'river of gold' in Guinea. This map, even today, holds a romantic allure to further exploration.

Prince Henry had one other great advantage. His brother, Prince Pedro, had collected much useful material in his years of travel at other European courts between 1419 and 1428. Most valuable of all his prizes was probably the copy of Marco Polo's account of his overland travels to China which was given to Prince Pedro in Venice together with a map 'which had all the parts of the earth described, whereby Prince Henry was much furthered'. Marco Polo's journey had been unique and at this period copies of his account, with all their relevance to the Indian Ocean and the lands of spices that lay beyond, were not widely disseminated.

Prince Henry's great contribution at Sagres to the theoretical side of the discoveries, and particularly to the cartography, was that he focused attention on these realistic maps and first-hand accounts rather than on the symmetrical and theological school exemplified by the Psalter and Hereford maps. He encouraged the careful plotting of coastlines, as practised in the Portolanos, rather than attempts to rationalize legends and find a place for accepted mythology in current geography.

One of the principal ways in which he demonstrated this inclination towards the scientific was by inviting a motley collection of scholars to Sagres, and by giving them employment and patronage there. Jaime Cresques, a Catalan Jew from Mallorca and the son of the famous Abraham Cresques *Magister Mappamundorum*, was persuaded to settle and work at Sagres; he became one of the leading cartographers of his age, helping Prince Henry to record and correlate the information brought back by his sea captains. The Mallorcan school of cartographers had already developed mathematical tables to assist in the measurement of distance at sea and these new methods were unhesitatingly applied at Sagres. The limits of their capabilities were recognized, however, even by themselves, and much information gained by the voyages down the African coast was never adequately processed at Sagres; there is a record of such material being sent in 1458 to Venice for mapmaking there, although doubtless the Portuguese were careful not to disclose too many of the secrets* of a coast which – they hoped – would open up a route to the spice islands which would be independent of the Venetian Republic's overland trade.

In fact, Prince Henry showed a marked disregard of race and creed in selecting his collaborators at Sagres. He installed numbers of Jews, since these had enjoyed a greater freedom than Christians to travel in the Islamic interior of Africa. (He also protected Jews at Tomar, where they were allowed their own synagogue almost in the shadow of the Templars' castle.) He had Moslem merchants and travellers at his court. Zurara particularly remarks upon the variety of nationalities to be encountered there; as well as Catalans, Jews and Arabs, there were Genoese and Venetians, Scandinavians and Germans, Berbers and Guineans. Zurara also records the awarding of gifts to 'Indians and Ethiopians', although he may have been indulging in some poetic fancy in this respect. Such a cosmopolitan collection of savants and travellers argues a frame of mind very different from the rigid crusading spirit of the participant in the Ceuta or Tangier expeditions.

*The Portuguese became increasingly secretive about their progress on the West African coast. When, later in the century, Pero d'Alemquer, the pilot of Bartolomeu Dias and Vasco da Gama, boasted at Court that he could take any ship – not only a caravel – to the Guinea coast and back, King João II rebuked him sharply and publicly for such idle boasting; the King explained to him later that he had done so to discourage foreign listeners from attempting such an enterprise.

And it was not only at Sagres that Prince Henry gave evidence of enquiring intellectual interests. He had held the title of Protector of the University of Lisbon almost since he reached manhood, and from 1431 onwards he had exercised this function with some interest and activity. As always, it is difficult to determine the exact degree of personal initiative and direction that lay behind the reforms and improvements that occurred during the years of Prince Henry's involvement, but it is surely possible to see his hand in some changes when one observes what those changes were.

First, the number of faculties was increased during his years of interest to include subjects related to his own preoccupations: arithmetic, geometry and astrology schools were set up. None of these was directly endowed by Prince Henry, who did however endow a chair of theology and remembered the theological lecturers in his will. Secondly, the material prosperity of the university was improved: the first university college was founded during these years, and the university acquired for the first time proper buildings of its own. Thirdly, and perhaps most characteristically, the discipline and morals of the students were brought under review: a new range of statutes was introduced which not only covered such minor sartorial details as the length of gowns which should be worn but also decreed that, 'No student shall keep in his quarters a horse, donkey, falcon or – on a permanent basis – a woman of loose morals'. Could it be that the ascetic Prince had a hand in the drafting of these statutes?

But however much or little he was involved in Lisbon, it is in the activities and atmosphere of Sagres where his claim lies to a Renaissance mind. It can be objected that all comparisons between medieval and Renaissance thought and attitudes are invalid. Indeed, Professor C. S. Lewis argued in his inaugural lecture as Professor of Medieval and Renaissance Poetry at Cambridge University in 1955, that the distinction between the two periods was so blurred as to be meaningless: all Renaissance attitudes were, he maintained, rooted in the medieval world. But this was surely a deliberately provocative overstatement of the case. For most of us the distinction is a real one and well illustrated by the difference between, on the one hand, Prince Henry's empirical activities at Sagres in the cause of exploration and discovery and, on the other, the medieval approach to the same questions as exemplified by the contemporary popular travelogues of Sir John Mandeville. While the latter – whose very existence is dubious and whose journeys were patently spurious – was regaling European readers with tales of the finding of Noah's Ark and of islands guarded by maidens in dragons' forms, Prince Henry was busying himself with lateen sails, astrolabes and serious charts. If the spirit of the Renaissance means anything, it is surely something very close to the rejection of religious superstition for determinable knowledge; in its most extreme form this sentiment finds expression in the lines attributed by Marlowe to Machiavelli in *The Jew of Malta*:

> I count Religion but a childish Toy,
> And hold there is no sinne but Ignorance.

Prince Henry had given more moderate expression to a similar thought over a century earlier at Sagres when he is reported to have written, 'knowledge is that from which all good arises'.

Chapter 8

Further into the Unknown

'Yet all experience is an arch where thro'
Gleams that untravell'd world, whose margin fades
For ever and for ever when I move.'

Lord Tennyson

With the traumas of the Tangier campaign, and the initial dispute about the regency arrangements on the death of King Duarte behind him, and with the added expertise in navigation and ship-building which the savants of Sagres had put at his disposal, Prince Henry was once more ready in 1441 to continue the progress of exploration down the West African coast.

In that year he sent out the first caravel under the command of Nuno Tristão, a knight of his household. Tristão had the usual orders to go further than his predecessors and to endeavour to bring back a native of the country for interrogation by Prince Henry. His ship followed hard on the heels of a more modest *barca*, which had been despatched to the Rio de Ouro, to collect the skins and oil of the sea-lions for which the area was already known, under command of Antão Gonçalves. This captain had been variously described as Prince Henry's Chamberlain and as his Master of the Wardrobe, and all his crew were recruited from the household of the Prince, whose sponsorship could not have been more direct.

When Tristão joined Gonçalves he found that the latter had already achieved the task allotted to the former: he had taken two captives. After loading his ship with sea-lion skins, Gonçalves had made a short excursion inland and, on his return, more by luck than judgement had found and captured a Tuareg man and a negro woman.

Tristão had been provided by Prince Henry with a bedouin

Arab from Morocco as an interpreter for just such an eventuality. It was a disappointment to both commanders to find that the captives spoke no Arabic. Tristão determined to try to capture others and the two captains organized a further raiding party in which, after a night march, they fell on a small encampment. Four natives were killed in the ensuing skirmish and ten were captured.

One of these prisoners was a most valuable prize. He was a minor chief from the Sanhaja Tuareg tribe (whom the Portuguese called Azanegues and who have been frequently and inaccurately described as Berbers). This man's name was Adahu and he had travelled fairly extensively, including to Morocco. To the delight of Tristão and Gonçalves, the interpreter discovered that he understood and spoke Arabic. Tristão decided that such a useful source of information should not be hazarded on further adventures but sent directly back to Prince Henry at Sagres. He therefore despatched Gonçalves homeward.

But before proceeding further down the coast himself, Tristão performed a significant ceremony. He formally knighted Gonçalves in recognition of his prowess in the capture of the natives. Here was chivalry of a strange kind indeed! A squalid skirmish was made the occasion for the winning of spurs, and a nocturnal scuffle in the sand against virtually unarmed savages was equated with noble feats of gallantry. Such an act indicates the determination of Tristão to treat these voyages as firmly within the crusading tradition. Whatever Prince Henry might feel of Renaissance sentiments and thought processes, his captains (at least at this stage) were trying to reconcile their actions with the framework of the medieval world. Their chronicler – Zurara – saw nothing strange in this: he, too, was totally absorbed in the illusion of chivalry as the mainspring of action.

Having performed his curious ceremony, Tristão sailed down the coast to perform another much more practical – and in reality much more significant – feat. He ran his light caravel aground and careened its bottom on a deserted beach. This was reported by the chronicler as evidence of how self-confident

Tristão felt in these strange waters, and on these inhospitable shores. No sea captain enjoys the sensation of being vulnerable upon the beach, and yet Tristão behaved 'as if he were in Lisbon roads'. Even more convincingly, it was evidence of the manoeuvrability, on the beach as well as on the sea, of the new, light caravel.

Tristão, his ship once more in good shape, proceeded a further fifty or more miles down the coast until he came to a headland of white cliffs jutting into the ocean. Beyond the point, a large bay opened out, which was clearly subjected to strong currents. Supplies were running low and the objectives of the expedition had been successfully achieved. Tristão named the headland *Cabo Branco* (Cape Blanco it has remained ever since) and turned his caravel for home and the welcome and rewards which Prince Henry reserved for those who had served his ambitions well.

Meanwhile, Prince Henry had been losing no time in questioning Adahu at Sagres. For the former it was the most stimulating and intellectually exciting exercise he had had for many years: here was an empirical method of checking the theories and principles expounded to him by his visiting savants. Not since his talks at Ceuta with the Count of Viana more than twenty years before had his imagination been so inflamed by news of the gold trade that came overland from the regions further south, of the great river (assumed to be a branch of the Nile) which flowed westward into the Atlantic, of the fabled city of Timbuktu where the caravan trails of the desert met the canoe trade of the rivers from the south. But possibly most intriguing of all to Prince Henry was the intimation that beyond the Saharan desert lay a green and fertile land, inhabited with heathens awaiting conversion to the true faith.

The medieval crusader and the Renaissance enquirer were equally stirred in Prince Henry by these revelations. His response illustrates both sides of his interests as well as his increasing awareness of commercial possibilities. He immediately despatched an envoy to the Pope with requests reflecting his varied concerns: he asked that spiritual jurisdiction might be granted to the Order of

Christ over all those lands to the south which he might discover, and that those who lost their lives in these voyages might be considered to have fallen on crusade; and at the same time he requested that a concession in perpetuity should be granted to the Crown of Portugal of the proceeds of those lands which might be discovered 'between Cape Bojador and the Indies'.

The mission was a total success. By a Bull of 1443 Pope Eugenius IV granted Prince Henry's requests. The Pope was probably glad enough to hear of the prospect of new conquests for Christianity at a time when the eastern fringes of Christendom were under such pressure from the Ottoman Turks. He additionally granted Prince Henry indulgences for the Church of Santa Maria da Africa which the latter had founded in Ceuta, thus no doubt consciously consolidating the Church's hold on that city at a moment when its surrender might still come under contemplation. These Papal benedictions and grants were to be confirmed by Eugenius' successor – Pope Nicholas V – and later in the century by Pope Sixtus IV. In fact, the Bull of 1443 was the first in a whole series of Papal pronouncements; those of 1452 (*Dum diversas*), 1455 (*Romanus Pontifex*) and 1456 (*Inter caetera*) in particular consolidated Portuguese positions in the newly discovered and shortly to be discovered regions, and also recognized explicitly Prince Henry's own decisive contribution to the progress already made both in the Atlantic islands and on the African coast.

These concessions from Rome were matched by equally significant ones from the Regent. Prince Pedro granted to his brother a charter allocating to Prince Henry the whole of that one-fifth share of the profits of the expeditions which would normally have been the prerogative of the Crown. He also gave Prince Henry a mandate by which all captains sailing down the African coast required the latter's permission: a monopoly had been established on the trade from these new shores. Whereas, previously, the successive expeditions had been a continual drain on the personal fortune of Prince Henry and on the funds of the Order of Christ, from now on, if an income could be derived, this could be set against such expenses. There was every inducement to maximize such income. A new weighting had been given to the commercial factor in relation to other motivating factors of crusading, missionary and exploratory zeal.

But Prince Henry's immediate concern was to deal fairly with Adahu. He had conceived the notion that this Tuareg, being a chief in his own country, was therefore to be treated with the courtesy and consideration reserved for a knight captured in battle. When Adahu had concluded telling the Prince all he could of the nature of the country from which he came, and of the stories he had heard of more distant parts, he expressed a strong desire to return to the Rio de Ouro; he found the Algarve too cold for him and the winds of Sagres a poor substitute for the warm air of the Sahara. He promised a ransom from his own people, not only for himself but for two of the other young captives who – Adahu assured the Portuguese – were also people of substance in their own land.

These considerations appear to have been uppermost in Prince Henry's mind when he fitted out a *barinel* for Antão Gonçalves in 1443. When Gonçalves eventually reached the Rio de Ouro, he – acting no doubt on Prince Henry's chivalric principles – released Adahu on parole, so that the latter might seek out his own people and arrange for his own ransom. Zurara approved this display of trust by one knight (albeit a newly created one) in another (albeit an imagined one). Adahu disappeared into the desert wearing a handsome suit of clothes, which Prince Henry had presented to him, as became a prisoner of war returning from one court to another.

Gonçalves waited for a week on his ship for the ransom party to arrive. It had been agreed that a dozen lesser captives at least should be exchanged for Adahu and there had been hopes of gold or other evidence of the riches of which he had spoken. Such hopes had almost evaporated when, on the eighth day, a Moor appeared riding a white camel and leading a small caravan of followers. He had no ransom for Adahu whose claims to knightly status appeared to have been ill-founded. But the Moor was prepared to barter for the two young captives whom Adahu had said would be ransomable. Eventually these were exchanged for ten negroes, some gold dust, a leather shield and – most exotic of all – a collection of ostrich eggs. These last were subsequently served to Prince Henry and somewhat improbably declared by the chronicler to have been found 'exceeding fresh' and a dish available 'to no other Christian prince'.

On this trip, Antão Gonçalves had had an unusual passenger. An Austrian nobleman named Balthazar, from the household of the Holy Roman Emperor, Frederick III, had won his spurs at the capture of Ceuta and had retained a great regard and respect for Prince Henry. He had heard the reports, now spreading across Europe, of the remarkable progress made by the Portuguese in penetrating the unknown waters of the Atlantic and the African coast. He had consequently resolved to go to Sagres and seek to relive the exciting achievements of his youth by volunteering for one of Prince Henry's expeditions. In particular, he announced on arrival at Sagres that he was anxious to experience a real storm at sea; he had heard that the Atlantic had tempests of a magnitude and splendour unknown to the enclosed waters of the Mediterranean. For Prince Henry, the Ceuta enterprise had always preserved an aura of romance, and all who had partaken in it shared in this aura. He willingly consented to Balthazar's request to join Gonçalves in the dangers and rigours of a long voyage in a *barinel*, though he must surely have wondered whether, if the medieval conventions could be strained to allow of a nobleman like Balthazar going on such an expedition, it was not possible for him too to do so. But his own princely status was different, and the convention too strong for him; he stayed at Sagres and sent Balthazar with Gonçalves. The visitor saw his tempest; in fact, both he and the whole ship's company were nearly lost in such violent storms that the *barinel* had to turn about and return to Lagos for repairs. But Balthazar, although it was nearly twenty years since Ceuta, had not lost his nerve. He sailed again with Gonçalves on the voyage with Adahu, and returned to Sagres to be honoured by Prince Henry and later to the Emperor's Court to recount great tales of all he had seen and heard.

The same year, 1443, Prince Henry also sent Nuno Tristão on a further expedition to sail – as always – further down the coast. He passed his own previous furthest point, Cape Blanco, and reached an island which was to be called Arguim and which was later to be of great significance. For the present the most remarkable thing about it was that two extraordinary creatures – which the Portuguese sailors took to be great birds – approached Tristão's caravel. On closer inspection they turned out to be canoes propelled by the legs of the negroes who travelled in them, who used their limbs as oars giving a curious flapping impression.

By now, attempted capture had become the natural reaction of Prince Henry's captains to the sight of natives. As soon as the occupants of the canoes were identified as human, Tristão set about trying to seize them; he caught fifteen, and sailed on down the coast. Further exploration revealed little remarkable apart from an island inhabited by herons. He returned with his prizes.

The next year, 1444, Prince Henry fitted out six caravels. Among their commanders was Gil Eannes who had first rounded Cape Bojador. They returned to the bay beyond Cape Blanco in which Arguim and other islands lay scattered, and they systematically set about raiding the mainland and the island for natives. Between them they captured two hundred, ranging in colour from the darkest black to the lighter shades of those who had admixtures of Arab or Berber blood. These captives were no longer specimens of new breeds brought home for the disinterested study of Prince Henry. They were a commercial commodity: the European slave trade had begun.

When the six caravels returned to Lagos in the Algarve it was decided to sell off their valuable cargo. Most of Portugal, and particularly the south, was underpopulated and additional labour was much sought after. The chronicler Zurara attended the sale of the slaves and gives the following account of what was clearly a memorable event:

On the 8th of August, 1444, early in the morning on account of the heat, the sailors landed the captives. When they were all mustered in the field outside the town they presented a remarkable spectacle. Some among them were tolerably light in colour, handsome, and well-proportioned; some slightly darker; others a degree lighter than mulattos, while several were as black as moles, and so hideous both in face and form as to suggest the idea that they were come from the lower regions. But what heart so hard as not to be touched with compassion at the sight of them! Some with downcast heads and faces bathed in tears as they looked at each other; others moaning sorrowfully, and fixing their eyes on heaven, uttered plaintive cries as if appealing for help to the Father of Nature. Others struck their faces with their hands, and threw themselves flat upon the ground. Others uttered a wailing chant, after the fashion of their country, and although their words were unintelligible, they spoke plainly enough the excess of their sorrow. But their anguish was at its height when the moment of distribution came, when of necessity children were separated from their parents, wives from their husbands, and brothers from brothers. Each was compelled to go where fate might send him. It was impossible to effect this separation without extreme pain. Fathers and sons, who had been ranged in opposite sides, would rush forward again towards each other with all their might. Mothers would clasp their infants in their arms, and throw themselves on the ground to cover them with their bodies, disregarding any injury to their own persons, so that they could prevent their children from being separated from them.

Besides the trouble thus caused by the captives, the crowds that had assembled to witness the distribution added to the confusion and distress of those who were charged with the separation of that weeping and wailing multitude. The Prince was there on a powerful horse, surrounded by his suite, and distributing his favours with the bearing of one who cared but little for amassing booty for himself. In fact he gave away on the spot the forty-six souls which fell to him as his fifth. It was evident that his principal booty lay in the accomplishment of his wish. To him in reality it was unspeakable satisfaction to contemplate the salvation of those souls, which but for him would have been for ever lost.

And certainly that thought of his was not a vain one, for as soon as those strangers learned our language they readily became Christians, and I have myself seen in the town of Lagos young men and women, the children and grandchildren of these captives, born in this country, as good and true Christians as those who had descended generation by generation from those who had been baptized in the commencement of the Christian dispensation. Nevertheless there was abundant tear-shedding when the final separation came, and each proprietor took possession of his lot. A father remained at Lagos, while the mother was taken to Lisbon and the child elsewhere. This second separation doubled their despair. However they were not long in becoming acquainted with the country, and in finding in it great abundance. They were far less obstinate in their creed than the other Moors, and readily adopted Christianity. They were treated with kindness, and no difference was made between them and the free-born servants of Portugal. Still more: those of tender age were taught trades and such as showed aptitude for managing their property were set free and married to women of the country, receiving a good dower just as if their masters had been their parents, or at least felt themselves bound to show this liberality in recognition of the good services they had received. Widow-ladies would treat the young captives that they had bought like their own daughters, and leave them legacies in their wills, so that they might afterwards marry well and be regarded absolutely as free women. Suffice it to say that I have never known one of these captives put in irons like other slaves, nor have I ever known one who did not become a Christian, or who was not treated with great kindness. I have often been invited by masters to the baptism or marriage of these strangers, and quite as much ceremony has been observed as if it were on behalf of a child or relation.

This incident has always worried Prince Henry's biographers. Zurara himself is clearly on the defensive throughout this account, labouring to maintain that the kindness subsequently afforded to the slaves excused the inhumanity of their initial treatment. Richard Henry Major, the great mid-nineteenth-century biographer (whose translation is quoted above to give something of its flavour) devoted a whole chapter to refuting accusations that Prince Henry had been responsible for starting the slave trade in general, for originating the deportation of African slaves in particular, and – indirectly – for the introduction of slavery into the New World. That Prince Henry was innocent of these charges is hardly a justification for – still less an explanation of – his actions. He could not, after all, be excused on the grounds that he was merely complying with the practices of his age. How, then, did this dedicated Christian crusader and (in many respects) enlightened thinker come to preside over what was the first slave market in Europe?

It was surely a supreme instance of the degree to which Prince Henry was the victim of that distortion of reality which was recognized in an earlier chapter to be one of the illusions of the chivalric age. He was forever living his own 'vision of a dream'. As we have seen, the encounters with primitive African tribesmen were presented to him – and by him – in terms of knightly escapades. We have seen how Nuno Tristão knighted Antão Gonçalves after the latter's capture of some natives. We have seen how a supposedly superior 'prisoner of war' was allowed to negotiate his own ransom. It followed – by Prince Henry's distorted chivalric logic – that the other Africans rounded up on marauding raids ashore and shipped back to Portugal were, somehow, common prisoners of war. Although Europeans did not sell each other's soldiery into slavery, the peoples of Islam did. There were precedents, therefore, for prisoners being treated in this way, even Christian precedents. When the Holy land was settled by the Crusaders, Moslem prisoners were frequently – if temporarily – used as slave labour on the estates of the King of Jerusalem. Others were employed more permanently as slaves in the households of the great Christian lords there. But this had not been in Europe itself and the precedents were tenuous ones.

The argument that only by bringing these natives back to Portugal as slave labour could their conversion to Christianity, and thus their salvation, be achieved was an even more tortuous one. The obvious method of converting the heathen to Christianity in substantial numbers was to convert their chiefs and rulers who would, in turn, ensure the compliance of their subjects to the new faith. This was to be the pattern for early Christian imperialism: even the conquistador Pizarro – whose claim to be considered a true Christian was far less well-founded than Prince Henry's – was to make at least a superficial attempt, in 1532, to convert Atahuallpa and his Inca following to Christianity before taking the decision to kidnap him and massacre his followers. Slave raiding diminished the prospects of establishing a dialogue – with a view to conversion – with native rulers. To argue that the best method of saving souls was to carry the bodies connected to those souls back to Europe, and then to sell those bodies into slavery, was palpably false. And yet Prince Henry, and Zurara, clearly convinced themselves that the conversion of their 'prisoners of war' was a principal justification for taking such prisoners.

Why did Prince Henry *wish* to justify such action? Surely here the answer lies in the economic realities of the ventures on which he was embarked. Commercial benefit had probably been low on the initial tally of objectives for the expeditions; indeed we have earlier examined these objectives and found this to be so. But now that the expeditions were multiplying, that their existence was sanctified by the Papal Bull of 1443, and that their profits could be used to finance fresh expeditions by the terms of Prince Pedro's concessions to his brother, there was every inducement to ensure that profits ensued. Other considerations apart, only in this way could a continual supply of fresh sea captains be found. As the scale of the voyages of discovery increased, it was no longer possible to draw exclusively – as in the past – from Prince Henry's household for their captains; a new class of adventurer had to be induced to sail to those distant waters. And while gold remained largely a chimera, slaves had become a concrete reality, so the supply of such adventurers increased. With each fresh headland passed, Prince Henry's fervour for discovery increased; with each fresh cargo of slaves, the finding of captains and crews became easier. The two processes – exploration and slave-trading – had become complementary. And the market for slaves in the underpopulated Algarve, seemed inexhaustible.

But the tragedy of the Lagos slave market was that Prince Henry had allowed his distorted chivalric ideals and his exploratory ambitions to override his Christian charity. As he rode up on his strong horse to supervise the division of African families and rejoice at the salvation of their souls, he apparently failed to observe the fundamentally un-Christian nature of the deed he did. He was paying the price for the glorious illusion he had fostered since – and even before – the Ceuta adventure. His military reputation had suffered at Tangier from the blindness induced by his strange and personal motivations; now his spiritual reputation was to be damaged in the same way.

Not for the first time in his life, the strains in Prince Henry's policies came into conflict with each other. One of the motives for permitting the slave trade to become established was that this would encourage his sea captains to undertake their voyages. Ironically, it also encouraged them to neglect those aspects of the voyages which were of most interest to Prince Henry. Gonçalo de Cintra, a young squire of the Prince's household, was despatched in 1445 (the year following the first slave market) with strict orders to proceed to the furthest part of the discovered coast without putting in anywhere or allowing himself to be deflected from his mission or exploration. But Gonçalo de Cintra found the prospect of making a quick profit, by some slaving on the way, to be irresistible. He landed in the island of Naar, near Cape Blanco, with some of his crew. His native interpreter escaped and may well have managed to warn the inhabitants of the vicinity. At all events, no Africans appeared and Gonçalo became ever more desperate to capture someone or – as he was reported by the chronicler to have put it – 'to perform some mighty deed'.

He therefore crossed various creeks and allowed himself to get separated by the incoming tide from his ship's boat. It was at this juncture that some two hundred natives appeared and fell on the twelve Portuguese seamen. A fierce fight ensued; five of Gonçalo's men eventually swam to safety; Gonçalo himself could not swim nor could three of the others; three more preferred to remain and fight to the last. Thus it was that seven lives eventually paid the price of Gonçalo's greed and the first casualties of Prince Henry's voyages were inflicted. Slaving was competing with discovery.

It must therefore have been with particular satisfaction that Prince Henry managed in the same year of 1445 to persuade another of his retinue – a squire called João Fernandes – to undertake a mission of research, devoid of any slaving implications, which would materially augment the knowledge available at Sagres about the hinterland of the newly-discovered coastline. The Renaissance side of Prince Henry's character longed for reliable data, for first-hand information from impeccable sources, rather than the customary mixture of fable and fantasy which could be gleaned from frightened captives.

João Fernandes volunteered for a unique service. He was prepared to be left on the Rio de Ouro to spend a whole winter exploring the interior as best he could. He had learnt some Arabic from Moorish prisoners and had a smattering of the Berber or Tuareg tongue. He made a rendez-vous with Antão Gonçalves, who undertook to pick him up in a caravel seven months later. According to Zurara, Fernandes made contact with the tribe to which belonged those young chiefs whom Gonçalves had brought to Portugal and back again. They stripped him of his clothes and remaining provisions, but then appear to have adopted him as one of themselves. He was invited to go on a long expedition by camel through the desert to the camp of a chief named Ahude Maymom; their water ran out and they had to steer by the stars and the flight of birds, as if they had been at sea. But, the journey completed, Fernandes was well received, fed on milk and eventually made his way back to the coast and his rendez-vous.

On return to Portugal, Fernandes was able to give Prince Henry information on many points of interest. He was able to confirm that the desert peoples among whom he had been living were of different tribes but all Moslems. He was able to recount the range of their camels, which covered fifty leagues in a day. He could describe the fauna of the desert; he had seen antelope and gazelle, ostriches and storks – the latter flying to more southerly climes. He had discovered the swallows from Portugal wintering in the warm Saharan sky. He had learnt to adopt the dress of the desert – the long, flowing robes of the Tuareg – and thus to survive the fierce sunshine of the day and the cruel cold of the night. All this was first-hand evidence.

But even more fascinating to Prince Henry must have been the overwhelming weight of reports which Fernandes brought back of a green and fertile land, not much further to the south, inhabited by the negroes whom the Arabs themselves enslaved and sold further north. This promised land beyond the desert (in fact, the basins of the Senegal and Gambia Rivers) was the source of the gold dust which Fernandes was shown and of the golden anklets which the richer Berber and Tuareg women wore; it was described to him – and by him – as the land of the Western Nile. Prince Henry now had even more credible evidence than that of the Tuareg chief Adahu to encourage further exploration. He also had, as will be seen shortly below, direct news from seaward of this area.

For the next two years a steady stream of caravels infested the region of Arguim Bay. In August 1445, a fleet of twenty-six ships sailed to the area to 'pacify' the islands in the Bay, some sailing there from Lisbon and others from Lagos. The accounts of these marauding expeditions defy even the chroniclers' abilities to glorify them; they abound in repetitive tales of a few Africans kidnapped in one place or a larger number seized somewhere else. 'It is a marvellous thing,' records Zurara, 'that as soon as one of these people is captured, he is prepared to point out to his captors not only other natives, but even his friends, wife and children, so they may also be taken.' Children were frequently seized without their parents, and Álvaro Fernandes, encountering a negro woman of such strength that she could not be dragged to the waiting boat, did not hesitate to carry off her two-year-old child in the (correct) expectation that the mother would follow. On another occasion, Zurara recounts that when the Portuguese had upset some native rafts 'the Christians passed amidst the rafts and chose above all the children, in order to carry off more of them in their boat . . . on numbering the Moors* whom they took during these two days, apart from a few who died, there were forty-eight . . . and they rendered thanks to God for giving them victory over the enemies of the Faith.'

The rapacity of the Portuguese sea captains had by now antagonized most of the inhabitants of this stretch of the coast and there are increasingly frequent mentions of resistance resulting in the loss of Portuguese lives too. Gonçalo Pacheco lost seven men and one of his boats, which was broken up by the natives to extract the nails from it; subsequent reports from the area revealed that the Portuguese casualties had been eaten, but Zurara explained that even this enormity was consistent with the code of honour of the region, where the liver and blood of an enemy would be devoured as an act of ritual revenge.

The Portuguese presented their own acts of vengeance against recalcitrant natives in Arguim Bay as being opportunities 'to gain honour'. A striking example of distortion of the chivalric code was provided on one occasion in 1445 when a punitive mission of this type had been proposed but presented problems to some of the Portuguese captains as it would have delayed their return to Portugal and their provisions were running short; they therefore proposed to throw half their captives overboard to save rations rather than 'lose the chance of gaining honour'. Happily, it seems that other ships arrived with fresh provisions in time to prevent this grotesque event.

But while the hunt for slaves became ever more frequent,

*In Zurara's vocabulary all Africans, including negroes, are 'Moors'; all the African coastline south of Cape Bojador is 'Guinea'; native chiefs are 'Dukes' or 'Knights', and the Senegal River is dubbed 'Nile'.

lucrative and squalid, some of Prince Henry's captains were still prepared to obey his injunctions to sail ever further down the coast. One such was Diniz Dias who in 1444 had reached Cape Verde, thus becoming the first of the Portuguese navigators to reach 'the Country of the Blacks'. (Zurara is confused as to whether he or Nuno Tristão was in fact the first.) Hitherto, all the negroes who had been captured – and the majority of the captives had not been negroes – had either been voluntarily outside their own homelands or had already been enslaved by Moors or Berbers and sold or exchanged. Now Dias was in their country; he had reached beyond the arid Saharan lands into the green, lush regions of which Prince Henry had already heard so much. His testimony was even more vivid than the reports which João Fernandes was bringing back from his sojourn in the desert.

Always a little further, was the persistent admonition from Sagres. Beyond Cape Verde, about sixty miles further south, another cape was discovered by Álvaro Fernandes and named the Cape of the Masts, because it was identifiable by tall palm trees, stripped of their fronds by a tropical tornado. At almost every major cape which the Portuguese explorers rounded from this time onwards they placed a commemorative column (known as a *padrão*). At first these were of wood, but later they were carved in stone in Portugal with the cross of the Order of Christ and the royal arms, and were carried out at considerable inconvenience in the caravels. The captains themselves showed great awareness of the significance of breaking new ground. Diogo Afonso, for instance, remarked after placing a huge wooden cross near Cape Blanco in 1445 that 'well might anyone of another country marvel who might chance to pass by that coast and see among the Moors such a symbol, if they did not know that our ships were sailing in that part of the world.'

It was Nuno Tristão who reached the mouth of the Gambia River in 1446. Of all Prince Henry's captains, Tristão was the one who most consistently reflected his master's intellectual curiosity; who shared most closely the Prince's Renaissance thirst for knowledge and capacity for experiment. It was he who had first

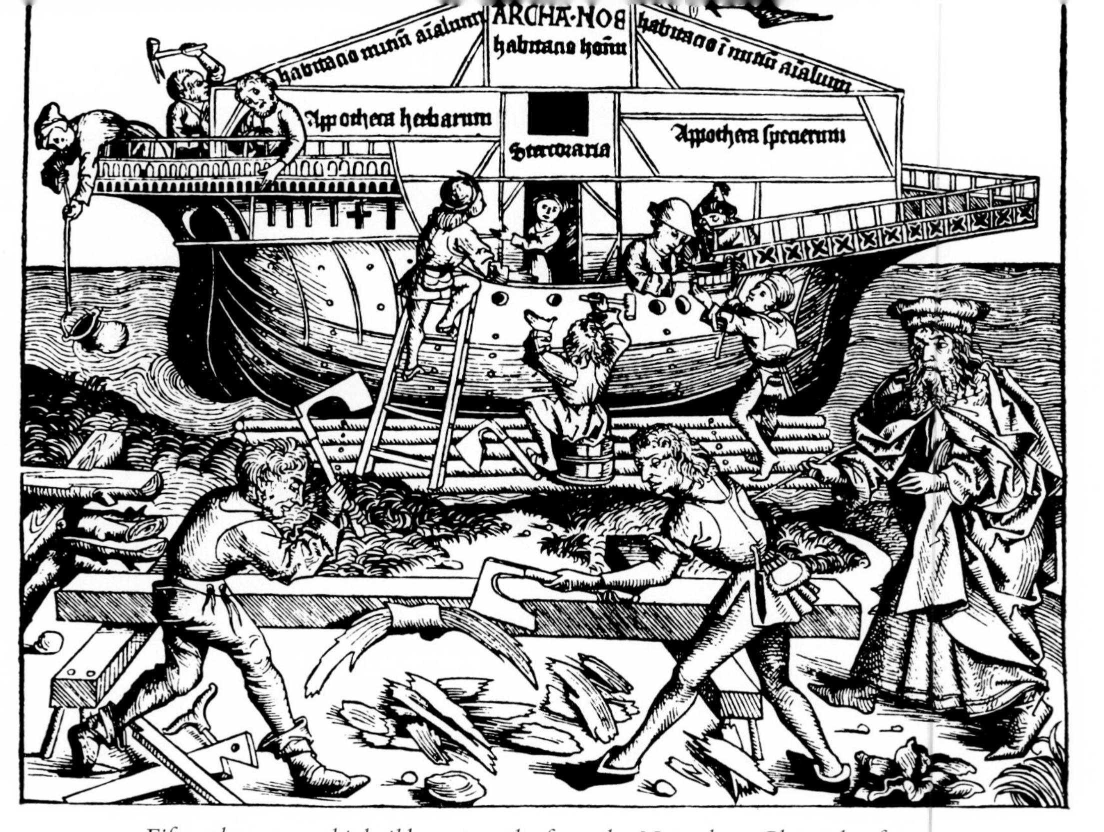

Fifteenth-century shipbuilders at work, from the Nuremberg Chronicle of 1493

The Catalan Map of 1375, from which Prince Henry would have learnt of the gold trade in the south Sahara

sighted Cape Blanco, who had first landed on Arguim Island, and who first commanded a caravel. Now this curiosity was to be his undoing.

Having reached the mouth of a great river, he dropped anchor and decided to explore upstream in two of the ships' rowing-boats with crews amounting to twenty-two seamen in all. He intended to seek a village on the river bank and learn what he could. The tide carried his boats rapidly upstream and the rain-forests closed around the small party.

It was now the turn of the intruders to be ambushed. Just as the Portuguese had heard reports of the natives of this area, so the inhabitants of the banks of the Gambia River had heard reports of the Portuguese landings further north. And they had not liked what they heard. From the moment when Tristão's boats entered the mouth of the Gambia they had been under observation from the natives hidden in the foliage on the banks. By the time the two small boats were some way up the river, the natives had managed to assemble twelve canoes containing about eighty warriors. These now fell upon the Portuguese, showering them with a hail of poisoned arrows from the canoes and from the banks. Tristão ordered a rapid retreat to the caravel, but even before he had regained the ship, four of the Portuguese had died of their poisoned wounds. Two more of the crew of the caravel received fatal wounds from arrows while endeavouring to raise the anchor: they cut the cable and put out to sea. Sixteen more of the sailors – including Tristão himself – were to die almost immediately of the poison.

This was disaster enough, but for the survivors it was only the beginning of their ordeal. The fifty-ton caravel was 1,500 miles from home and far further south than any other of Prince Henry's ships, and only five souls remained alive on board. This would have been desperate indeed even had the five been able-bodied and experienced seamen with a navigator among them. But such was not the case: they comprised a wounded common sailor with no knowledge of navigation, an African boy captive and two Portuguese boys who had been attendants on two of the knights who had been killed. Happily the fifth of the survivors (the boys had all been too young and inexperienced to be taken on the river adventure) was a young 'groom of the chamber' to Prince Henry called Aires Tinoco. This youth had not only learnt to read and write but, by attendance on the Prince at Sagres, had listened to much talk of navigation and the arts of sailing by the stars. Prince Henry, as a liberal patron of intelligence, had clearly encouraged the boy's interest. This was now to pay a dividend. For two months Tinoco directed the labours of the other lads. He held to a course north by east, keeping watch on the Pole star by night and the sun by day. They remained out of sight of land. The sails of the caravel were too heavy for them to operate efficiently, but somehow they kept going. At last they saw another ship and feared it would be a Moorish corsair. However, not only were their fears unjustified but they found from the captain of the ship they had sighted that they were approaching the coast of the Algarve. They made harbour at Lagos and told their experiences to Prince Henry.

The Prince had always rewarded his captains handsomely for breaking new ground. Tristão had done this on his last expedition, as on so many previous ones – but now he was not there to be rewarded. Prince Henry did what to us must seem commonplace, but what to his contemporaries was both remarkable and remarked upon: he pensioned the widows and dependants of all Tristão's associates who had lost their lives on the fatal voyage. Where no false concepts of chivalry intervened, Prince Henry could still be a charitable and compassionate man.

One more set-back was to mar this chapter of discoveries. Among the distinguished visitors to Sagres was a Danish nobleman called variously Vallarte or Abelhart. Prince Henry allowed him to go on a voyage of discovery under the tutelage of Fernando Afonso. They proceeded slowly down the coast and it was six months before they reached the island of Palma near Cape Verde. While some negotiations were on hand between Abelhart and the natives, by which the former was offering a tent 'which would shelter thirty men but be light enough for one to carry' in exchange for an elephant 'alive or dead', Abelhart ventured ashore and was ambushed on the beach. Only one of his companions managed to swim back to the caravel, the others being either slain or carried off as prisoners. Disconcerting reports reached Prince Henry for many years to come of four Christian hostages held at an African camp far in the interior, but no word from Abelhart was ever heard again.

Prince Henry was now almost at the height of his fame. King Henry VI of England had sent to create him a Knight of the Garter in 1442; like his father, Prince Henry cherished such visible evidence of chivalric distinction. He was acclaimed as he rode through the Portuguese countryside, not only in the Algarve but on his frequent visits to Tomar, Viseu and Lisbon. Recruits flocked to Sagres and Lagos to man his caravels and great travellers from elsewhere in Europe paid pilgrimage to his retreat. He had even given an African lion as a present to a foreign admirer. We know from Zurara that up to 1446 over fifty caravels had reached the Guinea coast (as he called the furthest point so far reached) and that by 1448 (the latest period covered by his Chronicle) 927 slaves had been brought back from this coast to Portugal. Prince Henry must have congratulated himself that the chivalric and the enquiring sides of his personality were for once in harmony. The inconsistency in his position appeared to have evaporated. But not for long. Another interlude in the discoveries was about to be forced on Prince Henry, and – as with the Tangier interlude – it was neither to be happy nor to reflect to his credit. But before that unhappy interlude of power struggles and domestic politics is recounted, it seems appropriate to examine in rather greater detail the economic backcloth against which the events both before and after this interlude were enacted.

Chapter 9

Counting the Cost

'The Lusitanian prince, who, Heaven-inspired,
To love of useful glory raised mankind
And in unbounded commerce mixed the world.'

James Thomson

A distinction can always be made between the motives of the protagonists in any story and the reasons why that story developed as it did. Some attention has already been paid to the motives of Prince Henry in connection with the discoveries, but there is a modern school of historiography that maintains that these motives were, in fact, almost irrelevant to the real causes of the discoveries: the latter being intractable economic forces. While not accepting this theory, it is necessary now to look at the whole economic background to the voyages of discovery and attempt an assessment of the part which such economic factors played in the events described.

It is indisputable that there was an urge for expansion in Portugal at the beginning of the fifteenth century. The wars with Castile were over but the nobility still wished to extend their estates; below them in the social scale, the class of new knights wished to establish estates for the first time. The newly emerging urban and mercantile class, who had supported João de Aviz to the throne and afterwards, were anxious for new markets. There was a requirement for fresh sources of supply of wheat and other cereals, and a sad shortage of gold in the royal exchequer.

There appeared to be three possible directions for such expansion. One was towards the Mediterranean. But the possessive attitude of Castile towards the Moorish Kingdom of Granada, and the likely opposition of other Christian monarchs to any adventures in that area, was a deterrent. It will be remembered that King João deterred Prince Henry from assaulting Gibraltar in 1418.

The second potential area of expansion was into the Moorish kingdoms of Morocco. Such expeditions were to appeal to Prince Henry throughout his life. For him, this appeal was largely that of a crusade: Ceuta (in 1415), Tangier (in 1437) and finally Alcácer Ceguer (in 1458) were occasions for chivalric enterprises against the Infidel. But there were other reasons too for these ventures. They were expected to provide the nobility with occasions to gain new estates for themselves, by the only acceptable method – that of conquest. They were further expected to provide a rich hinterland of wheat plains. It was hoped to capture the northern end of the gold routes across the Sahara and thus terminate the chronic shortage of gold in Portugal which was undermining the currency. It was also hoped that they would diminish the Moorish capacity to harass shipping along the coast of the Algarve. After the capture of Ceuta, subsequent expeditions had the additional incentive of attempting to capture territory which would end the isolation of Ceuta.

The third potential area for expansion was into the unknown. Even this was divisible into two separate directions: either into the Atlantic in search of islands which might provide useful crops (cereals and sugar), or along the African coast in search of gold – and later slaves, spices and other commodities. (The term 'spices' covered a wide range of medicinal drugs, perfumes, cosmetics and seasoning; but various forms of pepper were the main items.) This direction had less immediate appeal to the nobility than did Morocco; they saw no prospect of conquering new territorial estates, and the grander nobility could not participate in nautical expeditions as they could in campaigns of conquest. Impecunious knights, however, might hope to establish modest estates for themselves on newly discovered islands. And certainly for the merchant classes such expeditions, with their prospect of new merchandise and fresh markets, were more appealing than military campaigning in North Africa.

Thus the unknown was to be the chosen direction of expansion. Prince Henry's early successes in the Atlantic helped to finance his subsequent African voyages, since Madeira proved a profitable proposition for settlement and development. Some authorities have cast doubt on the primacy of Prince Henry's rôle in this, but it is relevant to note that the Papal Bull *Romanus Pontifex* of 1455, which granted important privileges to the Portuguese Crown, specifically recognized and paid tribute to Prince Henry's colonization of Madeira as one of the justifications for such privileges. The fact that it was he who imported the special vines from Crete and the sugar beet and other plants that were to be so successful on the island, is also evidence of his key role in the colonizing process. From the end of the 1420s, Madeira was exporting profitable produce and generating an income which could contribute towards the costs of the voyages down the African coast.

In fact, the development of Madeira was a model which could have been followed with advantage elsewhere. The Governor distributed among the colonizers the title to plots of land, each plot limited in size to that which it was considered possible for the owner to cultivate within a period of ten years. At the end of that period, the concession would be extended provided that the colonizing farmer undertook not only to cultivate the extension but to maintain the upkeep of the original parcel of land. In this way an incentive was established for good husbandry. All the wells and water courses were kept in public ownership, as were the water-mills. In these circumstances, Madeira prospered to the extent of producing 3,000 hogshead of wheat alone by 1446, as well as valuable quantities of wine, sugar, meat, honey, and – particularly relevant – timber for shipbuilding.

Nor was Madeira unique in being a profitable island. The Azores – having poorer soil – were developed much more slowly; cattle and sheep were imported to the islands in the early 1430s and allowed to proliferate there in freedom. Even by the end of that decade, there was no permanent Portuguese colony. But by 1443 we learn of the islands producing beef and some wheat for export to the mainland and by 1447 these and other products were in regular supply from the islands (as is evidenced by a letter from Prince Pedro giving permission for fish, wood, vegetables, wines and other products of the Azores to be unloaded without let or hindrance in any port in Portugal). Prince Henry later introduced sugar cane here too and a trade was established with Bristol. Even the Canary Islands contributed something towards the Portuguese economy and the costs of further discoveries: it is recorded that in 1438 Prince Henry was receiving a percentage of the profits from gums and dyes exported from Salvage Island in that group.

But the profits from the island discoveries were neither large enough nor did they come soon enough to contribute more than a modest fraction of the overall costs of those voyages down the African coast, which were ultimately the favoured outlet for Portuguese expansion in Prince Henry's lifetime. The bulk of the money had to come from elsewhere. Eventually, as will be seen, the voyages became highly profitable and, indeed, attracted merchant adventurers not only from Portugal but from the Italian maritime cities; but for the first twenty years in which Prince Henry was sending ships down the African coast, the profits were minimal and the costs prodigious. Where did the money come from?

In the early years, it came basically from two sources. One was Prince Henry's own fortune. It is impossible to gauge the exact extent of this but he always had a number of estates, offices and monopolies which provided him with funds. Admittedly, some of the most lucrative of these were not available to him between 1423 and 1443 when he was most in need of funds, but it is none the less illuminating to list here his various more obvious forms of income, and it must be remembered that his titles carried estates and Prince Henry was always assiduous in ensuring that the estates produced profits. He was Duke of Viseu and Lord of Covilhã; he was also lord of a long list of small places including Vila de Gouveia, Lagos, Alvor and the Berlengas islands; he was Governor of the Province of the Algarve; he was Governor of

Ceuta; he held the monopoly for tunny fishing in the Algarve; he held the monopoly for making and selling soap throughout Portugal; he had exemption from the requirement to pay a one-fifth share to the Crown of any booty taken by his ships; he had a monopoly of authorizing Portuguese voyages to the Canary Islands; he held the monopoly of coral fishing between Capes St Vincent and Spartel; he had the right to receive one fifth of the catch of various sorts of sea fish; he had a monopoly of all commerce between Madeira and the Azores, and between Madeira and the African coast; he alone could authorize voyages down the African coast, and on these voyages – as on all the others for which his authorization was required – he received a generous proportion of the profits as the price of such permission.

However, when this formidable list is considered, it emerges that most of Prince Henry's income derived from sources which were only developed, largely by his own energies, in the latter part of his life. His monopolies and his fifth shares became valuable assets after the African coast was being opened up to trade; they did not help to finance the early years of – often fruitless – exploration. So in addition to the income from his estates, some other major source was required.

It was here that Prince Henry's control of the Military Order of Christ was so useful to his purposes. He was not, in fact, Grand Master of the Order (as has been frequently maintained) but only because this title would have involved him in vows of chastity and poverty; the former would have been acceptable enough to the ascetic Prince, but the latter would have been quite incompatible with the substantial range of material interests enumerated above. He carried the title of Governor of the Order, and as such appears to have had an effective capacity for the deployment of the Order's funds.

And the funds of this Order were incomparable. It has been recounted in an earlier chapter how the Military Order of Christ came to be established as the successor to the Portuguese branch of the Knights Templar, when the Templars were ruthlessly put down throughout the rest of Europe. What has not been recounted is that the Portuguese branch of the Templars was among the richest of all. The mighty Cistercian monastery of Alcobaça (seventy miles due north of Lisbon) was protected from overland marauders be a screen of Templar castles, the most dramatically situated of which was Almourol,* built on a rocky island in the middle of the Tagus River, but the chief of which was that at Tomar. The Order had always been well endowed in Portugal, as successive kings had recognized their indebtedness to the Templars for their part in finally evicting the Moors from Portugal several centuries before the final eviction from Spain.

At a period when wealth – particularly in Portugal – was generally in the form of land, the particular advantage of this Order as a paymaster was that so much of its capital was apparently held in liquid assets; in effect it was available as 'risk capital'.

There were even strong rumours, the truth of which has never been established, that there was quite exceptional treasure at Tomar. These rumours were based on a theory that when Jacques de Molay, the last Grand Master of the Templars in Paris, was arrested on charges of blasphemy and sodomy on that black Friday the 13th of October 1307, he managed to send some of his personal treasure to the safety of Portugal. Since contemporary accounts of his wealth include a description of his arriving in Paris with 150,000 gold florins and ten horse-loads of silver, it is likely that any treasure he did manage to send would have been of appreciable value. Indeed the hideous torture inflicted on the Grand Master and his fellow Templars in Paris may – not inconceivably – have been connected with attempts to locate the missing treasure. A strange fabric of superstition arose at Tomar around the mythical golden ornaments, jewelry, chalices and *objets d'art*.

Whether any of this extraneous wealth existed or not, there were funds in plenty at Tomar and Prince Henry, as Governor of the Order, did not find it either difficult or improper to devote

*Commanded by that Gonçalo Velho Cabral who had first reported the sighting of the Azores to Prince Henry.

these in considerable measure towards fitting out ships to explore both the Atlantic and the African coast. The use of the funds of a religious order necessitated that the voyages of discovery should be presented as having a high religious purpose: the discomfiture of Islam and the conversion of the heathen to Christianity. It has been suggested by some historians that the need to present a convincing case to the Order of Christ for the use of its funds was why Prince Henry listed these motives among his objectives. But everything we know of his upbringing and temperament indicates that crusading zeal was among his genuine motivations, and if his methods of converting the heathen were somewhat curious (as was seen in the last chapter) it seems unlikely that Prince Henry was aware of any disparity between his own objectives and those of the Order into whose coffers he was dipping.

By 1443 the profits of plundering the West African coast had become sufficiently apparent for missions to be sent to Rome to secure Papal backing for the Portuguese position on the coast. Although the immediate return in gold was disappointing, the introduction of black slaves to Portugal had begun to be a profitable trade which was to grow steadily during the following decades. Soon spices and such *recherché* objects as ostrich eggs, dyes and unusual skins were to be added to the list of wares brought home. As the sugar estates developed in Madeira and in the Algarve (the first Portuguese sugar plantation and mill were set up there in 1404), so the requirement for slaves also increased; these slaves also provided a form of capital in metropolitan Portugal, where there was a considerable scarcity both of negotiable capital and of labour. Populations had been declining all over Europe during the previous century, largely as a result of the Black Death and numerous other deadly epidemics, and – Portugal being no exception – the influx of manpower was important.

From this date onwards we do not need to examine Prince Henry's own resources, nor those of the Order of Christ, to find the money used to equip the ever-increasing number of caravels and of whole fleets which were sailing down the African coast. They were mainly financed by private enterprise. Citizens of Lagos, such as Lançarote, equipped their own expeditions and looked for very attractive profits. Professor Magalhães Godinho has calculated that such profits were seldom less than 100 per cent and were not infrequently as high as 700 per cent of the sum laid out. From 1445 until 1463 there was a special trading house at Lagos which dealt exclusively with goods returning from the Arguim trading station on the African coast, and while the chroniclers were to estimate (as has been shown) that nearly one thousand slaves had been brought back to Lagos during the years 1441 to 1448, some subsequent economic historians have put the figure at double that number. These represented profits which must have seemed to justify the twenty years of investment in voyages down a barren coastline.

And to a nation whose seamen had traditionally been fishermen, there was always the attraction of fresh fishing fields in the newly discovered waters. Whatever their motives for reaching strange waters, the Portuguese were never too busy to find time to cast a line.

Ironically, the expeditions against the Moors in Morocco, which had initially appeared to be an alternative to voyages of exploration down the African coast, now became complementary to them. As Prince Henry's captains discovered that it was easier and more profitable to purchase slaves and other goods than to seize them by force, so it became increasingly necessary to have goods to offer in payment and exchange. Most of the goods the Africans coveted were only available to the Portuguese in Morocco: surplus agricultural produce and textiles (the Africans had an apparently insatiable desire for woollen cloaks) were chief among these, but such objects as Cowrie shells and coloured beads – although originating from the Levant or Venice – were often more easily obtained by Portuguese merchants in North Africa than elsewhere.

It has been argued that the shift in emphasis in the voyages of discovery, the shift towards maximizing profitability, had little to do with Prince Henry. Professor Magalhães Godinho has suggested that only about one third of the voyages down the African coast were undertaken on the direct initiative of Prince Henry, the other two thirds being at the initiative of the knights, squires and merchants involved, or at the instigation of the Regent – Prince Pedro. Indeed, the same authority has pointed out that, in fact, the most rapid progress down the coast was achieved during the eight years of Prince Pedro's Regency (1439–1447) when 198 leagues of new coastline were revealed, as opposed to only 94 leagues in the subsequent twelve years. Other scholars too have drawn attention to the fact that it was Prince Pedro who consistently championed the urban merchant classes; he had been their candidate for the regency, and he had consistently favoured exploration rather than the type of military adventure in Morocco which appealed to the nobility. Significance too is attached to the fact that it was Prince Pedro who had travelled widely, in Eastern Europe and elsewhere, and it was he who had brought back copies of the writings of Marco Polo and others. Surely, the exponents of this school of thought maintain, it was therefore Prince Pedro who should be hailed as the patron of the Portuguese discoveries and new commercial frontiers?

Prince Pedro's role was undoubtedly substantial, and much greater than had been recognized until recently. But to suggest that he overshadowed Prince Henry seems perverse. It was Prince Henry who moved his centre of operations to Lagos and Sagres to be close to his sea captains and to enable him personally to supervise the fitting out of expeditions. It was Prince Henry who granted permission to knights and merchants to undertake voyages even when in the later years he was no longer the main paymaster or motivator of all those voyages. It was he who persuaded a new class of adventurer to open up commercial negotiations where previously his captains had been intent on raiding and marauding. It was he whom the chroniclers eulogized as the great architect of Portuguese navigation (and this they did not only in the years following Prince Pedro's disgrace and fall, but consistently thereafter).

Above all, in the context of the finances of the expeditions,

Prince Henry's contribution stands supreme. It was he who found the funds, from his own resources and those of the Order of which he was Governor, for the early unproductive ventures. And when the voyages themselves had become profitable, it was he who continued to devote his fortune to encouraging captains and cartographers, squires and sailmakers, astronomers and travellers to settle at Sagres and to labour in the cause he had made his own. His curious Court was one from which few went away empty-handed. With all his estates and monopolies, Prince Henry might have been expected to die rich; in fact, he died in debt. Although economic reasons were an important factor in the discoveries, and, indeed, an increasingly important factor in Prince Henry's own calculations, money was never to become a mainspring of his life. He had incentives enough already.

Chapter 10

Power Struggles and Politics

'Plots, true or false, are necessary things,
To raise up commonwealths, and ruin Kings.'

John Dryden

To understand the political crisis which demanded Prince Henry's attention in 1448 it is necessary to look back over the whole period of the Regency, since the death of King Duarte ten years earlier.

The King's will had been an unusual one. Apart from his resolve to use his fortune to ransom his brother Fernando, he had surprised the Court by naming his widow, Leonora, as sole Regent during the minority of his son Afonso. Since the boy was only six years old at the time of his father's death, it seemed likely to be a lengthy Regency. And if the Court was surprised, the Council of State and the burghers of Lisbon were positively indignant. The late King had brothers of proven ability and integrity surviving him; surely – it was argued – they, rather than a woman and a foreigner, should be entrusted with the management of the realm. Prince Pedro, as the eldest, was the obvious choice.

Not everyone was dismayed at the prospect of Queen Leonora as Regent. Many of the aristocracy were potentially over-mighty subjects, who knew that Prince Pedro would restrain their rapacity more effectively than the Queen. Chief among these was the ageing Count of Barcelos, the bastard half-brother of the late King and of the royal princes Pedro, Henry and João. The sixty-year-old Count had always felt a grudge against his legitimate relations which had been aggravated by the conferment of dukedoms on Prince Pedro and Prince Henry; now he was remarried to a young and ambitious sister of the Archbishop of Lisbon who urged him to exert his influence at his brothers' expense. The Archbishop was himself related to Queen Leonora and also inclined to support her claims to the Regency.

It was into this already strained atmosphere that Prince Henry arrived from the Algarve. Both sides vied for his support. Neither succeeded in winning it, and he was therefore invited to act as an arbiter and suggest a solution to the dilemma. The document which he produced was a lengthy memorandum showing a diplomatic sensitivity and a totally different side of his personality from that displayed in the memorandum he had written before the Tangier campaign. Whereas earlier he had propounded a series of moral propositions reminiscent of a medieval divine, now he proposed a practical compromise full of the spirit of the new age. Perhaps he was reacting to the traumatic experience of the Tangier *débâcle*.

Prince Henry proposed that the Queen should be sole guardian of her children, including the royal heir. All matters pertaining to the Court and the royal household would be exclusively in her hands. She would bear the title of Regent which King Duarte had intended for her. But Prince Pedro would be granted the new title of Defender of the Realm and all decrees would need his approval as well as that of the Regent. Where the Regent and the Defender of the Realm could not agree, the matter would be referred for advice to Prince João and Prince Henry. The Cortes would be convened not less than once a year; a Council would be set up to form a court of appeal in the event of any serious deadlock. It was an ingenious solution, perhaps too ingenious to be workable.

The Queen was encouraged by her faction to resist any power-sharing; Prince Pedro, on his side, was sceptical about whether he could work with the Queen. The burghers of Lisbon were adamant, however, that the Queen was not to be allowed to rule alone, and so the uneasy partnership was initiated, with the blessing of the Cortes. Prince Henry withdrew to Sagres. The next two years were to be uneasy ones. The Queen struggled with her state duties through a pregnancy culminating in the birth of a posthumous daughter to King Duarte. But friction mounted rather than grew less with the attempt to govern the country in tandem. The citizens of Lisbon remained vocal in their support of Prince Pedro and hounded the Archbishop from the city; riots ensued. The Queen attempted to stimulate suspicion of Prince Pedro; she even wrote to advise Prince Henry that his brother intended to arrest him. In this she badly over-reached herself: Prince Henry showed the letter to Pedro. The two brothers had complete confidence in each other at this stage. It became clear that matters could not carry on in this way.

Eventually the Cortes appointed Prince Pedro as sole Regent. Only after Prince Henry's personal intervention, and with him as her escort, would the Queen consent to bring the young King to the Cortes at all. Her malice towards Pedro and her neurotic behaviour prejudiced the Cortes against her; they declared that Prince Pedro should be charged with the education of the young King and his younger brother, who should share his household. Prince Pedro invited the Queen to do so also, since he rightly estimated that the royal children were too young to be separated from their mother. However, Queen Leonora declined, preferring to be parted from her children than to endure residing in conditions which she considered to be humiliating. The Queen bade a dramatic nocturnal farewell to her two sons and embarked on a life of self-imposed exile, first in other parts of Portugal and later in Castile (her native country). Prince Henry, for his part, returned to Sagres, convinced that he could safely devote himself to his nautical pursuits; and this he did for the next seven years as has been described in a previous chapter.

Shortly before the end of that time, the problems of the Regency flared up again. In the year 1446 the young Afonso reached the age of fourteen and officially came of age; he was solemnly installed as King at a ceremony at the palace at Lisbon. As King Afonso V of Portugal he received the Rod of Office from the hand of the Regent in the presence of the representatives of all the Estates of the Realm. The young monarch was so overwhelmed by his new responsibilities that he implored Prince Pedro to remain as Regent for the present and to continue to rule the country as he had been doing for the previous years. With some misgivings – to which he alluded in an elliptical conversation with Prince Henry at Coimbra – Prince Pedro continued in office. But the Count of Barcelos, who had recently been created Duke of Braganza by Prince Pedro in a belated effort to assuage his bitterness, was intent on manipulating the young King for his own purposes. He therefore set about turning the King's mind against the Regent.

He had one great advantage in this task. After Afonso's mother – Queen Leonora – had gone into exile, she had continued to plot against the Regent. When eventually she tired both of this futile activity and of living abroad, she had written to Prince Pedro asking him to allow her to return to Portugal, not as Regent or Queen but 'merely as his younger sister'. Pedro had immediately consented; but it was not to be. The unhappy Queen died in 1445 before she could set out on the journey – apparently from poison. In her death, as in her life, she was to generate hatred and suspicion.

It was to these events that the Duke of Braganza now turned the mind of the young King. It was suggested to him that Prince Pedro would have been embarrassed by the return of the King's mother, Queen Leonora, and that he therefore had her poisoned to maintain his own paramount influence over her son. The distressing parting from his mother must have remained a sensitive memory to King Afonso and the Duke's slanderous suggestions were not dismissed with the contempt which – in view of Prince Pedro's honourable record – they should have been. Instead, he informed the Regent that, despite his earlier request that the latter should continue in office, he now wished to assume the full powers of sovereignty himself.

Prince Pedro did not demur, merely suggesting that to avoid confusion and an impression of indecision, the King's assumption of powers should be postponed until his marriage shortly after his fifteenth birthday. The bride was Prince Pedro's own daughter Isabel, but the choice had been King Afonso's personal one; he had resisted adamantly all those (including his own brother and the Duke of Braganza) who had at one time or another tried to deter him from this alliance with his uncle the Regent, and he was to cherish his wife with love and care throughout the whole of her short life.

Now that the King was reigning in his own right, the Braganza faction determined that the influence of Prince Pedro should be removed as far as possible from the Court. His nominees were dismissed from royal appointments and every effort was made to deny the ex-Regent personal access to the young King. Eventually the Count of Ourém, one of the Duke of Braganza's sons, persuaded the King that, while his uncle and father-in-law was at hand, the King would have no real independence; King Afonso determined to ask Prince Pedro to withdraw from the Court. But before the request could be made, Prince Pedro, who had learnt what was afoot, pre-empted the King by requesting leave to retire to his private estates at Coimbra.

He was not allowed to enjoy his retirement there unmolested. The Braganzas continued to plot against him. It was put about that he had been responsible for the deaths of his elder brother the late King Duarte, of the King's widow Queen Leonora, and even of his younger brother Prince João, who had died shortly before of a fever. The sheer affrontery of such charges was astounding. The circumstances of King Duarte's death ten years before had been well known, and there had never been the slightest previous suggestion of any but natural causes. Prince João had been a staunch supporter of Prince Pedro up to the moment of the former's death. There was not the slightest evidence of Prince Pedro's involvement in the death of Queen Leonora; poisoned she might have been, but in another country and at a time of rapprochement with Pedro. Castilian motives could have been more easily adduced than Portuguese ones for her murder.

On no one did these spurious and malevolent charges have a greater effect than on the one surviving brother of the accused:

Prince Henry heard of them in the Algarve and, immediately laying aside his nautical activities, he rode northwards across the plains of the Alentejo to Lisbon to protest at the slanders against his brother. Prince Henry was coolly received at Court. His own close association with Prince Pedro, to whom he owed his commercial concessions, weakened his testimony in his brother's support. Among the many others who had been befriended by Prince Pedro during the Regency and who had prospered under his rule, none appeared prepared to take up his cause now that he was in royal disfavour and that his enemies were dominating the Court. If Prince Pedro's reputation were to be defended, his champion must come from elsewhere.

Prince Henry understood this. Doubtless he recalled his mother's dying injunction to her sons to stand together; she had compared them to a quiverful of arrows which could be snapped separately but never broken together. Doubtless, too, this memory awakened in Prince Henry's mind the high chivalric mood of the Ceuta adventure. For a problem of honour he must have sought a knightly answer. He wrote for help to the one man whose reputation in this field stood paramount, to the man on whom the spiritual mantle of the 'Holy Constable' – Nun' Álvares – had fallen. He wrote to Álvaro Vaz de Almada.

Álvaro Vaz had a European reputation for heroism. In the year of Ceuta, he was one of that happy band of brothers who had fought with King Henry V of England at Agincourt. He had been created a Knight of the Garter in recognition, not of his rank or capacity for useful alliance, but of his courage and audacity. King Charles VII of France had also recognized these qualities and had created him Count of Avranches. The Holy Roman Emperor had bestowed special favours on him. He had fought beside Prince Henry at Tangier, and his dash and valour on that occasion had been one of the few bright features in a dismal campaign. Surely, Prince Henry must have reasoned, if the honour of his brother Pedro was impugned, this was the man to speak in his defence. King Afonso could not fail to heed such an advocate.

Álvaro Vaz responded to the summons by travelling immediately from Ceuta, where he was performing garrison duties, to Lisbon. The Braganza faction recognized the danger to their position which his advocacy represented. There was speculation at Court that the Duke would trump up charges against him. Álvaro Vaz presented himself, dressed in the full regalia of his honours, before the Council of State and challenged his detractors to support their accusations in personal combat. Such gestures appealed to the imagination of the fifteen-year-old King and must also have touched a sympathetic chord in Prince Henry's memories of the feats of the Knights Templar. The Duke of Braganza persuaded the Court to move to Cintra, where the King could be insulated from the influence of Álvaro Vaz. The latter went north to Coimbra to express his sympathy with Prince Pedro. Prince Henry, feeling no doubt that his brother's honour had been vindicated by the support of that arbiter of honour Álvaro Vaz, retired again to Sagres.

The action taken had been quite inadequate to meet the gravity of the danger. Prince Henry had underestimated the extent of the Braganzas' hostility, had misinterpreted the nature of their threat to his brother, and had responded in an inappropriately formalistic way. It was not Prince Pedro's honour which required defending, it was his independence and his life. Evidence of faith in his integrity by honourable men was no answer to the Machiavellian plots being hatched against him. (Although Niccolo Machiavelli was not to be born for another twenty years, the Portuguese court was already beginning to show some of those features whose indentification and codification he was to undertake in Renaissance Italy.)

The Braganza faction now shifted their tactics from vilification to provocation. They endeavoured to trap Prince Pedro into taking some action which could be presented as evidence of his disloyalty to the King. They engineered the removal of his son from the post of Constable of the Realm. They insisted that he stop fortifying his castles at Coimbra and Montemor, although there was no evidence that he was so doing. They required that he should surrender the arms which had been lodged at Coimbra some years previously when a threat was feared from Castile. They tampered with his correspondence with the King, in which – among other things – he sought to justify holding onto the arms as a means of defending himself if his enemies forced him to do so.

Prince Henry continued to urge patience on Prince Pedro. At all costs any act which could result in a confrontation was to be avoided. The young King would eventually see reason. Honourable conduct would finally triumph, Prince Henry was convinced.

When all these minor provocations failed to rouse Prince Pedro to some act of indiscretion that could be presented as treasonous to the King, the Count of Ourém and his father the Duke of Braganza hatched a more blatant plot. The Duke was on his estates in the North, while his son was at Court. The Count of Ourém persuaded the King to request the Duke of Braganza to come to Court, and it was arranged between the father and son that the Duke should interpret this royal command as authorizing him to march his armed retainers southwards through the lands of Prince Pedro in the centre of the country (around Coimbra) without the Prince's leave. Some clash seemed inevitable and the Braganzas were confident that the King would take their side.

Prince Pedro remonstrated with the Duke to no avail. Prince Henry sent a message from Tomar to his brother, urging him to do nothing rash. Álvaro Vaz meanwhile advised Prince Pedro in an exactly contrary sense: he should resist the Duke's forces with his own forces. Reluctantly Prince Pedro agreed that enough was enough: he confronted the Duke of Braganza's entourage. The Duke promptly withdrew, leading his armed men round Prince Pedro's lands by a circuitous route through the snow-covered Serra da Estrela. Prince Pedro rejected Álvaro Vaz's advice to pursue his enemies and remained thankful that bloodshed had been avoided. The elderly Duke succumbed to the rigours of the mountains in winter, and only arrived at Santarem to join the Court much later, having suffered badly in health from his adventure. As predicted, he enjoyed the King's sympathy for his self-inflicted sufferings.

It was surprising that the Queen could not have exercised more influence on her father's behalf. Queens of Portugal had been far from passive consorts: Queen Philippa had been a power in the land during the planning of the Ceuta expedition, and Queen Leonora had been instrumental in launching the Tangier expedition. Neither Queen had been more dearly loved by her royal spouse than was Queen Isabel. The difference was that Isabel was still a child and unable to appreciate or mobilize the influence that might have been hers. She did however make some endeavours. When she discovered that the King had been persuaded to march against her father to punish him for his resistance to the Duke of Braganza, she wrote to Prince Pedro warning him that only three options were to be left open to him: prison, banishment or death. Encouraged by Álvaro Vaz, Prince Pedro unhesitatingly chose the last, and prepared to sell his life dearly by marching south and insisting by force of arms on an audience with the King.

The Queen's alarm was now total. She begged the King to relent towards her father and this he agreed to do, if Prince Pedro would write to him in suitably contrite terms. Prince Pedro unhappily spoilt the effect of his 'penitent' letter by writing simultaneously to his daughter the Queen saying, in effect, that he had written to the King in the required terms to please her rather than because there was any justification for demanding such a letter. There is some confusion as to whether the offence was caused by the King reading the letter to the Queen or by passages in the letter to himself, but – at all events – the good effect was undone and the King's preparations to subdue his uncle, father-in-law and mentor – Prince Pedro – went ahead.

A final emissary from Prince Pedro to the King – the Prior of Aveiro – was denied admittance to the royal presence. With Álvaro Vaz at his side, Prince Pedro set off with 1,000 cavalry and 5,000 infantry to march on Lisbon; he expected to be intercepted by the King's forces from Santarem which he knew to be 30,000 strong. It was a desperate move. After spending a night en route in prayer at the Abbey of Batalha, where his father King João I lay buried and where his own tomb awaited him, he encountered the royal armies at Alfarrobeira on the 20th of May 1449.

It seems that to the last Prince Pedro hoped to be able to reason with the King. Certainly he must have realized that a battle would be fatal to him and his supporters. Olivier de la Marche, a Burgundian nobleman who wrote a contemporary account of the meeting based on reports of those who had been present, described how Prince Pedro made his deployments for a parley rather than for a battle. But as was almost inevitable with the two armies facing each other, fighting broke out and Prince Pedro was himself almost the first casualty, falling with a crossbow bolt through his light armour. His head was severed from his body by a soldier of the King's army who later claimed that he should be knighted for the deed. Álvaro Vaz suppressed the news of the Prince's death as long as he could, and himself later fell exhausted to be hacked to pieces by the royal soldiers who had overwhelmed him. Prince Pedro's supporters were quickly routed. It was a sad day for Portugal, for the House of Aviz and for that concept of unity and chivalry in which Prince Pedro and Prince Henry had been nurtured and towards which, in their different ways, they had so long striven.

What responsibility had Prince Henry for these unhappy events, and how can his behaviour in the affair be reconciled with his character? Some of the facts are uncomfortable for his reputation. He alone had the stature and reputation to stand up to the machinations of the Duke of Braganza and his family, and yet his efforts to counter their calumnies on his brother were futile in their effect and somewhat less than vigorous in their execution. Prince Henry does not seem to have asserted himself at Court with the energy which he had displayed earlier when urging other matters close to his heart. Prince Henry had had his way with his father over the Ceuta enterprise, and with his eldest brother over the Tangier expedition. He had also resisted all the pressures of the Court and his family to swap the unfortunate Fernando for Ceuta. He had brow-beaten his sea captains into venturing into unknown and perilous waters. He had won his concessions from the Pope for Portugal and for the Order of Christ. All his contemporaries agree that his force of character and will-power were almost irresistible. And yet he had failed to persuade a fifteen-year-old boy to behave with a modicum of justice or compassion towards his own uncle and father-in-law. Can Prince Henry really have had his heart in the task?

Before attempting to answer this, one should perhaps consider what claim Prince Pedro had on his brother's support. The two Princes had always been the closest among the brothers. Nearly of an age and neither of them in line for the throne, they had fought and received their dukedoms together at Ceuta. Both were Knights of the Garter and respected abroad. Prince Pedro's regard for his brother was vividly illustrated by the fact that in 1445 he had invited Prince Henry to Coimbra to dub his eldest son, who had become Constable of the Realm, as a knight; no greater compliment of brotherly trust and chivalric regard could have been paid. Prince Pedro had consistently shared and encouraged Prince Henry's interests in navigation and in the university. It was Prince Pedro who had granted to Prince Henry the commerical concessions, normally reserved to the Crown, arising from the trade with the newly discovered lands. It was to Prince Henry that Pedro naturally turned in his hour of need, and yet when the crisis came Prince Henry was not at his side, like Álvaro Vaz, but at the King's Court at Santarem surrounded by Prince Pedro's enemies and detractors.

Historians have taken very different views of Prince Henry's conduct, usually either condemning or excusing, but seldom explaining it. The nearly contemporary chronicler Ruy de Pina clearly was unsympathetic to Prince Henry and blamed him in some measure for the unhappy outcome of events. The distinguished Portuguese historian, Oliveira Martins, writing at the end of the nineteenth century, is even more vitriolic in his condemnation of Prince Henry's motives and actions. He points out that Prince Henry's advice from Tomar – urging caution and restraint – was evasive, and although he promised to discuss matters more fully when he joined Prince Pedro, in the event he never went to join him, but remained supervising the improvements to the Convent of Christ. He stresses how much Prince Pedro looked to his one remaining brother for help and moral support, and how he looked in vain. He blames Prince Henry for his total absorption in his African projects and implies he had a callous disregard for his brother's predicament. Others have gone further, and accused Prince Henry of actually desiring the destruction of his brother since he might have calculated that Pedro's removal from the scene would leave him better able to manipulate King Afonso V to his own purposes. Such critics point out that it was Afonso who was to acquire the appellation of 'the African' and that Prince Henry might have already recognized in this young monarch his most sympathetic ally and patron.

A different school of historians have as avidly defended Prince Henry's behaviour. The father of these was Zurara, who had finished his chronicle before the events described in this chapter, but refers to them none the less. Zurara recognized that Prince Henry had been beside the King and not Prince Pedro during the final stages of the tragedy, and indeed on the field of Alfarrobeira itself, but he said that those who blamed Prince Henry for this did not know all the circumstances, and he undertook to reveal these in his later writings; unfortunately, the promised explanation was never written, or – if written – has been lost to posterity. Most of the laudatory and uncritical biographies that have been written in Portuguese and other languages over the intervening five centuries have argued that Prince Henry did what he could and was not responsible for the fact that his efforts failed. Mr Ernle Bradford, in an excellent modern account* of Prince Henry's seafaring achievements, points out that his advice to Prince Pedro was basically sound: if Pedro had stayed on his own lands at Coimbra the worst that could have happened would have been prolonged sieges of his castles, in the course of which

***Southward the Caravels*.

some accommodation might have been reached with the King. He also points out that after the Battle of Alfarrobeira Prince Henry requested leave to go to Ceuta, presumably to escape from the atmosphere of the Court; but Prince Henry did not persist in his request and quickly returned to his habitual and familiar surroundings at Sagres.

If the accusations are too sharp, the defence is somewhat less than convincing. Perhaps the explanation of this curious episode in Prince Henry's life, like the explanation of so much else, lies in the dichotomy in his make-up. His instinct was to reach back to his upbringing and seek a simple chivalric solution to unpleasant problems: if his brother's honour was impugned, then the knight-errant Álvaro Vaz would come to the rescue. This was his 'native hue of resolution'. But when the problem revealed itself as one which could no longer be explained in these terms, when – in fact – it was clearly not a problem of honour at all, but of power politics and Machiavellian intrigues, then his chivalric upbringing failed him. He fell back on the only other comparable component of his personality – his methodical and enquiring mind – only to find it 'sicklied o'er with the pale cast of thought'. The medieval knight in his make-up had provided too insensitive and conventional a response; the Renaissance thinker produced no response at all. The rough world of fifteenth-century politics equated neither with the courtly conventions of his youth, nor with the serious and studious pursuits of his middle age. In his brother's hour of need, Prince Henry was found wanting, but through a failure of experience rather than a flaw of character. The immediate victim had been Prince Pedro; the later victim was Prince Henry's own self-respect and reputation.

Chapter 11

Raiding Gives Way to Trading

'Shy traffickers, the dark Iberians come:
And on the beach undid his corded bales.'

Matthew Arnold

Once more Prince Henry was in retreat at Sagres and once more, where others would have been broken by the humiliation of his experiences, he merely redoubled the energies he applied to the tasks he had set himself.

Almost the first of his endeavours after returning from Alfarrobeira was a final assault on the Canary Islands. The secure anchorages provided by these barren islands en route to the West African coast had long been envied by Prince Henry. Their ownership was anomalous and some explanation of their history is here required. A Norman nobleman – Jean de Bethencourt – had captured the islands at the very beginning of the fifteenth century and had, to a considerable extent, colonized some of them. He had sought the protection of the King of Castile who, in turn, had given Bethencourt some help in subjugating the islands. The latter had brought settlers from both Castile and France and had spread his control from Lanzarote to Fuerteventura and Hierro. With the Pope's blessing he had installed a Castilian bishop on the islands. Then, in 1406, Jean de Bethencourt retired to France and left his nephew Maciot de Bethencourt as governor-general of the Canaries; effectively, he handed over the islands, with a total of some one hundred and fifty Christian settlers, to Maciot. The position of the uncolonized islands – Gomera, Palma, Teneriffe and Grand Canary – was even less defined; the Bethencourts and the Castilian Crown considered however that they had rights over the whole group.

In 1414 the maladministration of Maciot de Bethencourt

caused the Castilian crown to send out Pedro Barba de Campos, Lord of Castro Forte, to regain control of the islands. Maciot ceded the islands to Barba, who subsequently sold them to another Spanish nobleman. Meanwhile Maciot sold them first to Prince Henry of Portugal and then sold them a second time to the same Spanish nobleman who considered he had acquired them from Barba. Meanwhile again, Jean de Bethencourt left the islands in his will to his brother and not to his nephew Maciot. The position could hardly have been more invitingly open to intervention.

In these circumstances it was not surprising that in 1424, having just established his settlements in Madeira, Prince Henry should have sent an expedition to seize the Canaries by force. But this had been considered by King João in the same light as Prince Henry's earlier designs on Gibraltar: it was altogether too provocative towards Castile and too expensive an undertaking. The temptation offered by these strategically placed islands, which were either maladministered in the name of Castile or awaiting colonization, remained to haunt Prince Henry.

There was a crisis in this respect in 1445. Some of Prince Henry's captains returning with a disappointing cargo from the African coast had made a detour to the Canaries. They had behaved despicably, taking advantage of the hospitality offered them on Gomera to launch a raiding expedition against the natives of Palma, they then returned and took prisoners from among the friendly inhabitants of Gomera itself. Prince Henry was horrified: this treachery offended both sides of his nature. On the one hand, it was a blatant breach of the chivalric code; on the other, it was damaging to the atmosphere of commercial confidence which Prince Henry was increasingly concerned to build up. He reprimanded the caravel captain responsible – João de Castilha – and gave gifts to the prisoners, who were safely returned as free men to their own island.

Since this incident occurred during the period of Prince Pedro's regency, Prince Henry had consulted his brother about regularizing the Portuguese connection with the Canaries. Prince Pedro had granted a charter to Prince Henry (one of several such) which forbade all Portuguese to visit the Canary Islands without Prince Henry's explicit consent; this applied whether their motives were belligerent or commercial. Prince Henry was also granted a one-fifth share of the profit from any goods imported from these islands. A lease of Lanzarote from Maciot de Bethencourt is claimed by some authorities also to have been granted to Prince Henry, in 1447; if this is so (and payments made thereafter suggest that it may be) then it is also the case that Maciot had leased the island to the Castilian Crown. The anomaly surrounding the question of rights in the Canaries continued.

So did the raiding. A few years later, a further effort was made under Prince Henry's direction to seize the islands. One raid on Palma resulted in the taking of twenty captives and their transportation to Portugal on a voyage during which the ships' victuals ran out and the Portuguese apparently gave serious consideration to eating their prisoners, but land was sighted before cannibalism had set in.

Some modern historians are inclined to view Prince Henry's attempts to seize and colonize the Canaries as a central theme in his life. The present author is not so inclined. It was surely only as a stepping-stone to the West African coast that Prince Henry was attracted to these controversial islands. To him they had neither the appeal of a crusade nor the lure of the totally unknown. Attention to them has been deferred to this point in Prince Henry's life because it was during the years after the fall of Prince Pedro that we find the best description of life in the Canaries. Its author is perhaps the most celebrated of all Prince Henry's captains, the Venetian explorer and trader – Cadamosto.

Alvise Cadamosto was an unabashed adventurer. It was by chance that he came into the service of Prince Henry; he was sailing in a Venetian galley from the Mediterranean to Holland and bad weather in the winter of 1454/55 detained him in the shelter of the Sagres promontory. Work was then under way – as it had been intermittently for so long – on the Vila do Infante at Sagres, and Prince Henry was staying nearby at Raposeira. One of his household went down to the shore to investigate the strange galley and took with him samples, as gifts for the ships' officers, of some of the produce of Madeira: sugar and 'dragon's blood' (a valuable reddish dye). Cadamosto's interest led to an audience with Prince Henry at which the young Venetian – he was twenty-two at the time – had his imagination so fired with tales of the riches and opportunities of the African coast that he resigned from his commission to Holland and volunteered to take a caravel to Africa for Prince Henry. The terms on which such service was accepted were strictly commercial: either Cadamosto could provide and provision his own ship, in which case Prince Henry would require only a quarter of the profits of the expedition, or Prince Henry would provide and provision the ship and take half the proceeds. Cadamosto chose the latter alternative and on the 22nd of March 1455 set sail, with a Portuguese sailing master, in a forty-five-ton caravel.

He sailed first to Madeira, where he was greatly impressed with the progress of the colonizers. He commented in his journal on the wine and the wheat, on the oats and the fruit, on the sugar and the timber. In the thirty-odd years of Portuguese settlement, under Prince Henry's direction, the island had been changed from a wild paradise to an irrigated haven with a population of over eight hundred. Everywhere there was evidence of scientific and constructive endeavour: an elaborate network of *levadas* (or narrow irrigation channels cut in the hillsides) covered some parts of the island already, and water-mills and saw-mills utilized the force of the mountain streams. This was a new world of progress for Cadamosto.

It was from Madeira that he sailed on to the Canaries, and immediately he noted the change from Prince Henry's world of development to the much simpler life of this Castilian settlement (there was no wine or corn, he remarked) and the totally primitive existence on the pagan islands of the group. Cadamosto was a keen observer and recorder: it is to him we owe a vivid picture of life on Teneriffe and Gomera. Of the former island he noted that the inhabitants generally went naked and painted their bodies.

The Templars' castle at Tomar. The twelfth-century rotunda was modelled on the Holy Sepulchre at Jerusalem and the knights attended the mass on horseback, standing in the ambulatory shown in the lower photograph

Details of the later (Manueline period) constructions at Tomar, depicting the cross of the Order of Christ and the ribbon of the Order of the Garter. Both had been associated with Prince Henry

The chiefs of the island exercised *droit de seigneur* and all the maidens were deflowered by them. We learn elsewhere that on Grand Canary the chiefs insisted on the maidens being fattened 'so that they bear large sons' and then slimmed again before they slept with them. On Gomera it seems that wives were common and offered as hospitality to guests. (Zurara had earlier commented that on this island 'their supreme happiness is in fornication', and that they 'ate things unclean and disgusting like rats, fleas, lice and ticks'.) On Teneriffe when a chief came of age one of his tribesmen would 'voluntarily' throw himself to his death in a ravine; there were even Christian witnesses to the gruesome ceremony. All this lurid detail was noted by the youthful Cadamosto, who then sailed further down the African coast.

When he reached Arguim Island he again noted the influence of Prince Henry. In the immediately preceding years a fortress had been built there and it had been established as a permanent trading station; in fact, this was the first not only of the Portuguese forts (which were to stretch in a chain as far as Macau within a little over the next century) but of all the other colonial forts and armed trading posts which were shortly to encircle the globe. Arguim was a hive of activity: by the time Cadamosto arrived it was linked up with caravan routes intersecting the desert and bringing in a steady flow of negro slaves and – to a lesser extent – gold. In return for these commodities, the Portuguese traded linen and woollen cloth (frequently made up into cloaks in the case of the latter), silver, carpets and grain. All those trading from Arguim Island required a licence from Prince Henry. Until the orderly development of this trade, the Portuguese had relied upon raiding to collect the slaves; for over ten years they had been sending out caravels to plunder the fishing villages around the bay of Arguim. Now all that was changing. No more was there talk of knightly skirmishes with enemies who would become 'prisoners of war'; from now on seven or eight hundred slaves would be purchased annually and shipped back to Portugal.

Cadamosto interrupted his voyage at Cape Blanco, whence he made a journey inland by camel. Here he found he was as fascinated with the sexual habits of the Tuareg tribesmen, as he had been with those of the natives of the Canary Islands. He noted that women were admired for the length of their breasts, and that young girls bound their breasts tightly in cords 'to make them grow so long they sometimes reached their navels'.

As a Venetian he must also have been particularly intrigued by two facts which he noted. The Arabs and Tuareg from the interior used Cowrie shells as currency; these had reached them from the Levant through Venice and then by the overland caravan routes. Also he noted that white pepper was a staple trading commodity among the Arabs and Tuareg and that it was already reaching Europe from this source; this was the beginning of that trade in spices which eventually – augmented by the richer spices of the orient – was to flow by the sea route to Europe and thus bypass Venice, the traditional clearing house for such trade.

From Cape Blanco he sailed on to the Senegal River and Cape Verde, where he again commented on the manners and morals of the inhabitants. The King of Senegal had never less than thirty wives, which he distributed around the villages through which he intended to travel. He then moved with his entourage, holding court in each village in turn until he had tired of the local wives or they became pregnant, when he would move on. The women wore nothing above the waist, he observed, but he commented that they washed three or four times a day. He deplored their eating habits, however, and noted that they were appalling liars. More significantly, he reported that they lived mostly by capturing their neighbours and selling them to Arab slave dealers; already the first stage of the slave trade – marauding – was no longer being undertaken by Prince Henry's followers, but by others.

Fifty miles further south from the Senegal River, Cadamosto reached the country of a chief called Budomel who had a reputation for being a good customer for Portuguese wares. Cadamosto sold him seven horses (it remains a mystery how he kept them alive on a small caravel in those temperatures for such a long trip) in exchange for a hundred slaves. Budomel appeared so well pleased with his bargain that he gave Cadamosto additionally a twelve-year-old negro girl whom the recipient found 'very pretty for a black girl' and sent her back to the caravel 'to serve in his cabin'. He also invited Cadamosto to make a fairly extended visit to the interior, which the latter, always anxious 'to see and hear new things', accepted. He thus became the first European to give any detailed account of a sojourn in Africa south of the Sahara.

Cadamosto spent a month at Budomel's village court. He was simultaneously shocked by the primitive nature of the chief's life and by the exaggerated awe in which he was held by his subjects. He describes a house with seven courtyards and the profound obeisances of those who approached the chief. He reveals his own fifteenth-century European outlook by commenting with disapproval on the fact that Budomel's prestige appeared to be based on his personal qualities – courage, strength, fair-mindedness and intelligence – and wealth, rather than on his inherited rank. And, as previously, he comments on the sexual predilections of his host, who had nine wives in each of his villages, each wife with six serving girls – with all of whom the chief slept. Possibly his prowess in these matters too added to the esteem in which he was held.

Cadamosto reveals how far he was a merchant and an explorer, rather than a missionary of Christ, most clearly when he recounts his rare proselytizing endeavours. He made a bald statement to Budomel, in the presence of the Mohammedan mullahs who frequented his court, that Christianity was the only true Faith; and he cheerfully records Budomel's reply to the effect that while Christians certainly seemed richer than Mohammedans in that part of the world, and therefore enjoying clearer evidence of divine favour, it was still the case that black men – being so poor – surely stood a better chance than Christians of salvation hereafter. Even religious conversations with Cadamosto rapidly seemed to turn towards profit and loss accounts. His conscience clear, Cadamosto quickly reverted to his main preoccupations of bartering goods and observing the curiosities of this new country.

And curiosities there were in plenty. He describes seeing elephants, lions, panthers, leopards, wolves and parrots of brilliant colours; these last he acquired in large numbers to resell on return. He particularly comments on the snakes, some extremely venomous and others large enough to swallow a whole goat. He recounts tales of superstition in countering the ill-effects of these serpents. If Cadamosto was continually amazed by what he saw, he gave equal amazement to his hosts. They rubbed his limbs to see if the whiteness would come off. They were convinced that a bagpipe which he had brought with him (presumably the Galician rather than the Scottish variety) was a live animal. They readily concluded that his caravel found its way across the sea by use of the painted eye on its prow.

Cadamosto may not have been the most fervent of missionaries but he had a commendable enthusiasm for giving practical instruction. He found that candles were unknown to his hosts and he set about teaching them how to make these for themselves from the wax of bees. (The self-indulgent natives had always previously eaten their honey in the comb.)

Finally he sailed on towards Cape Verde and, after joining up with two other caravels – one belonging to a squire of Prince Henry's household and the other to a Genoese merchant adventurer – decided to explore the region of the Gambia River, where Nuno Tristão had met his death. On the way there they landed one of the interpreters (all of whom were captured Africans, trained and provided by Prince Henry for the purpose) to parley with the natives on the shore; the Portuguese on their caravels had the humiliation of seeing their envoy killed by his interlocutors on the beach. Undaunted they pressed on and, when they reached the mouth of the Gambia River itself, decided to venture upstream in their caravels. Previously such expeditions had been made in the ships' boats, but Cadamosto and his companions felt sufficiently confident of their ability to manoeuvre their craft in a confined space to risk taking them between the wooded banks of a river known to be infested by unfriendly inhabitants.

It was very nearly a repetition of the earlier disaster. The natives shadowed the intruders in their canoes. First the boats scouting ahead, then the leading caravel and finally all the caravels were surrounded by hostile canoes. The Portuguese held their fire as long as they could, but when a shower of poisoned arrows was loosed in their direction they riposted with crossbow fire and – eventually – by noisily firing the ships' bombards. Although these were not aimed at the canoes, the loud explosion of such artillery was sufficient to drive off the attackers.

The three caravels made fast to each other and got their interpreters to hail the shore. They explained that they came to trade and not to capture prisoners. They even sent further interpreters ashore to try to convince the Gambians of the truth of these assertions, which were indeed the fact. But the ill-fame of Portuguese conduct in the past and further north had preceded them; the effects of a decade of ruthless marauding and slave-raiding could not be obliterated by a few verbal assurances. Fear of the white man had resulted in myths even more terrifying than the reality: the natives of the Gambia River really believed that the Portuguese ate the Africans whom they bought. No dialogue was possible.

Cadamosto and his companions withdrew to the open sea. The captains wished to sail yet further down the coast; surely, they argued, the lands of gold could not now be far off. But the seamen insisted that the vessels be turned towards home. They had already collected much merchandise; they had been away for almost a year and there were mutinous murmurings at the prospect of a longer separation from their families. Cadamosto noted that at this latitude the night sky was revealing fresh stars; from his notes it is clear that he had identified the Southern Cross. His first voyage was successfully concluded.

The dates of Cadamosto's second voyage are still controversial. He wrote that he sailed again 'in the following year', which would have been in 1457, and he claims to have discovered the Cape Verde Islands on that voyage. But this account is full of discrepancies and inaccuracies; it seems likely that he deliberately antedated his second voyage and that, in fact, this did not take place in the lifetime of Prince Henry. At all events, the account of that voyage continues the tale of trading south of Cape Verde in the so-called 'Kingdom of Gambia'. The gold was less available than Cadamosto had hoped but a high price was paid for trinkets. Confidence in the Portuguese in this area was restored: the rivers were crowded with boats 'like the Rhône, near Lyons' – Cadamosto observed – as they tried to bargain with the visitors. It was here that he first encountered hippopotomi – 'horse-fish' as he called them – and he records that he took home some elephant flesh for Prince Henry to taste (which is either evidence, or a deliberately misleading statement, dating the voyage during the Prince's lifetime). Eventually, they sailed so far south that their interpreters were no longer able to communicate with the natives of the region. They turned back only after reaching another great river estuary which, with a certain lack of imagination, they named the Rio Grande, but which was probably the Geba River (on which now stands the town of Bissau, for many years the capital of Portuguese Guinea).

One voyage which certainly took place around the year 1457, however, was that of Diogo Gomes, a 'faithful servant' of Prince Henry. He sailed in a caravel almost directly to the furthest point so far discovered, the Geba River, and noting the strong currents encountered thereafter turned about; returning to the Gambia he traded and made friends with the natives on the right bank of the river. They led him as far up the river as the town of Cantor, the highest point navigable by a caravel. Here he encountered face to face the merchants from Timbuktu and came into direct contact with the overland gold trade. He heard tell of the King of Kukia who was 'lord of all the gold mines on the right side of the river of Cantor' and who had a block of gold outside the door of his palace which took twenty men to move and to which he tethered his horse. It was here that Diogo Gomes also heard of a notable battle between two chieftains of the interior – Semanagu and Sambegeny – which had taken place shortly before; when on his return Diogo Gomes reported this news to

Prince Henry, the latter informed him that he had had intelligence of the battle two months previously through an Arab merchant operating out of Oran on the North African coast.

On his return journey down the Gambia River, Diogo Gomes managed to make peace first with one chief and then with another. He sent ahead an Indian interpreter called Jacob (had Prince Henry hoped he might indeed reach the Indies on this voyage?) to negotiate, and finally he managed to make friendly contact with Nomimansa, the chief who had been responsible for the death of Nuno Tristão and the earlier attacks on Cadamosto; he was, in fact, the main enemy of the Portuguese in this area. So great was Diogo Gomes' success with Nomimansa that the latter listened to his account of Christianity and promptly told his Moslem adviser to leave his court within three days. He asked that Prince Henry should send out a priest to baptize him 'Henrique' – after Gomes' patron. (This Prince Henry did as soon as was practicable.) He also asked for sporting falcons, for livestock and for two Portuguese master builders to help him construct more solid dwellings. He even presented Gomes with the anchor which Nuno Tristão's men had cut loose as they made their escape. This was different indeed from the showers of poisoned arrows that had greeted earlier visitors. The main problem now, as in the years ahead, was ill-health; both Diogo Gomes and Cadamosto lost many of their crew through sickness on this coast.

Finally, Diogo Gomes mollified the most vicious of all the chiefs on the Cape Verde section of the coast – a certain Beseguiche. This warrior was rash enough to go out in a canoe to investigate the caravel and accept an invitation on board. He was recognized by Diogo Gomes' interpreter who privately informed his master of the identity of the guest. Diogo Gomes first gently questioned the visitors as to why their chief should be so hostile to the well-meaning Portuguese, and then, when the visitors had taken their leave and were rowing away from the caravel, called after Beseguiche by name to inform him that he had known his identity all along and had spared him as an example of the goodwill which he wished to demonstrate.

Diogo Gomes was later to claim to be the first to have discovered the Cape Verde Islands when he was sent on a subsequent mission, but whether this was after or before the death of Prince Henry (of which he was a witness, as will be seen) is not clear. At all events he did not get the credit, which appears to have mostly gone to António de Noli, one of his companions, who arrived home earlier than Gomes and successfully claimed the captaincy of the Island of Santiago. It seems, however, that there can be no doubt that these islands were discovered in Prince Henry's lifetime and were among those 'bequeathed' by him on his death.

One thing emerged strongly from these voyages of Cadamosto and Diogo Gomes: they were practical men intent on maximizing their profit from their trips, both in terms of commercial gain and in terms of useful knowledge discovered. With their advent, the age of plundering was over; no longer were the natives of the coast or the islands hunted as a military or sporting pursuit; no longer was fighting the natural form of contact with the inhabitants. It has been found that slaves could be more easily bought than captured. Gold could only be procured by learning where it was to be found and such knowledge presupposed a dialogue with the inhabitants and the establishment of confidence. Other valuable commodities were found to be available to those who knew of their existence: spices, parrots, ostrich eggs. Diplomacy was necessary to undo the harm of the early ravaging raids; familiarity was necessary to disarm the fear engendered by earlier visitors.

Prince Henry, from his central clearing house of information at Sagres, was the first to realize all this. He saw in men like Cadamosto and Diogo Gomes ready instruments of the new policy, a policy that was orientated towards economics. This was not because Prince Henry had renounced his earlier interests in favour of money-making; it was because he recognized that only by keeping up the profitability of the voyages could their continuation be assured. If undiscovered waters had no profits to compare in value with routine trading at Arguim, then those waters would remain unvisited. He therefore continually urged on his captains that they should create the right conditions for trade. For instance, they should not fire first; their consequent reluctance to do so had contributed to the perilous situation of Cadamosto's expedition in the Gambia River. He was sharp in his condemnation of gratuitous raiding, such as that carried out in the Canary Islands by João de Castilha.

Prince Henry had not failed to note from the reports of all those of his captains who sailed beyond Cape Verde that the coastline inclined south-eastwards, rather than south-westwards, from that point. He hoped and believed that this indicated that the southernmost tip of Africa was approaching and that soon his caravels would be sailing into the Indian Ocean and directly to those lands of oriental silk and spice of which Marco Polo had written and from which a steady traffic was already reaching Europe through the Middle East and the Venetian Republic. He knew that casual marauding would have no future in these more sophisticated regions and that his captains must now learn to practise the skills of the merchant. The fact that the African coast was again to turn southwards after the Gulf of Guinea, and that there were thirty years of further expeditions before Bartolomeu Dias was to round the Cape of Good Hope in 1486, was never known to Prince Henry.

On the other hand, the depth of his knowlege about the Sahara and the whole interior of Islamic Africa was steadily increasing; he now was able to collate information brought to him by his captains with other information gleaned from different directions. By the time of Diogo Gomes' expedition, Prince Henry had reached the age of sixty-three; he had already acquired the reputation of an elderly sage.

As the accounts came in to Prince Henry from his prosperous settlements in Madeira and elsewhere; as the fort at Arguim consolidated its military and commercial position; as the flow of caravels with rich cargoes continued to arrive at Lagos; as the fund of applied knowledge available at Sagres became ever more evident – it must have seemed to many that Prince Henry had forever resolved the contradictions in his approach to life, that from now on he was committed to the sober, studious, enquiring and pragmatic values that had contrasted so sharply with his chivalric impetuousness. He was to prove them wrong.

Chapter 12

The Final Crusade

'Death closes all: but something ere the end,
Some deed of noble note, may yet be done.'

Lord Tennyson

The year 1453 was a traumatic one for Europe and for Christendom. For half a century the Ottoman Turks had been rolling across the Anatolian plains and the calls for help from the Emperor Constantine Palaeologus at Constantinople had been becoming ever more shrill. In Gibbon's words, 'the generosity of the Christian princes was cold and tardy'; it was a classic case of sending too little and too late. At the beginning of April 1453 a supply convoy of five ships (four of them Genoese) managed with great daring to reach the beleaguered garrison at Constantinople, but – although the Sultan's admiral was punished with one thousand strokes of a golden rod for allowing this – the reinforcements were inadequate to provide resistance for more than a few weeks. In a daring move, the Sultan transported his fleet ten miles overland to the upper harbour where it was beyond the reach of the Christian ships and able to harass the weakest section of the city walls. On the 29th of May the final and fatal assault took place.

When eventually Constantinople had fallen to the Sultan Mohammed II and the great cathedral church of Santa Sophia had been desecrated and converted into a mosque, then, and only then, did Pope Calixtus III appeal to the Christian monarchs of Europe to mobilize for a crusade to restore the Eastern Christian capital to the true faith. It was an emotive appeal (legend has it that the priest bearing the Host had disappeared miraculously into the walls of Santa Sophia and would reappear only when the city was freed) but it fell largely on deaf ears. Those kings who had remained unmoved by the final calls for help from the garrison of Constantinople were less likely to be galvanized now by a message from a Pope who was, in any case, known to be corrupt. Even in the Italian peninsula, where the encroachment of the hordes of Islam was a real and nearby menace, no move was made to answer the Pontiff's call.

Only on the extreme western fringes of Europe was there any positive reaction. King Afonso V of Portugal responded by offering to mobilize an army of 12,000 men to fight the Infidel. He minted a new gold coin – the cruzado, which carried the crusaders' cross – from African gold, to pay for the expenses of the army. But even the youthful King quickly realized that an army of such a diminutive size, if unsupported by larger armies from other Christian countries, could not possibly sustain a successful campaign at the other end of Europe.

It was at this juncture that Prince Henry, who was in close contact with his nephew the King, began to urge that the army which could not be deployed at Constantinople should be allowed to fight a campaign against the neighbouring Infidel, in Morocco. The young King Afonso shared his uncle's enthusiasm for a trial of strength with the King of Fez. Having raised a crusading army he was anxious to find an arena in which to deploy it.

There were other more practical reasons too for launching a further campaign in North Africa. Ceuta in isolation was a permanent problem; it had no supply lines and was subjected to constantly recurring attacks. A firmer foothold on the Moroccan coast would contribute towards the security of the Algarve as a whole, and diminish the threat from Moorish privateers; conversely, it would provide better facilities for the Portuguese privateers to harass Moorish shipping in the straits. A successful campaign in Morocco would secure more agricultural produce from that country for use not only in Ceuta, but also in metropolitan Portugal and for barter on the Guinea coast. Moroccan conquests could give Portugal access to the textiles of that country, which were also of increasing value to her for trade in Africa further south.

It was arranged that the King and his cousins should collect fleets in Lisbon and Oporto and that Prince Henry would gather a fleet in the Algarve. The three would join up as an armada to attack Alcácer Ceguer. This small Moorish coastal town lying between Ceuta and Tangier had been chosen as the objective for the assault only after much debate. The richer Moroccan town of Safim would have been a much more profitable objective, but it was so far from Ceuta that it would have been little use as a support for the latter, and very difficult to relieve if it had itself been surrounded. Tangier was also considered; it would have complemented Ceuta from the point of view of their mutual defence, and would have been easy to reinforce from the Algarve. But the memory of the disaster of 1437 was too vivid; the ghosts of that ill-fated expedition could not be laid; Tangier was rejected as an objective. The advantages of Alcácer Ceguer were that it was a provocation to the Portuguese already, being much used as a base for Moorish piracy. It was also a centre for textiles. It was conveniently close to Ceuta. And lastly, and perhaps most temptingly, it was not very strong or well-fortified.

In fact, some such further Portuguese invasion into Morocco had long been contemplated. A Bull issued by Pope Eugenius IV in 1443 had conceded to the Order of Christ the whole valley of Angera, Tetuao and Alcácer Ceguer 'when freed from the infidels'. The Portuguese had been biding their time to make good the conditions of the Bull. It has even been suggested that King Afonso's response to the later Papal call to avenge the fall of Constantinople was no more than a cover for raising an army to campaign in Morocco. Be that as it may, it was Alcácer Ceguer which was now to attract the full attentions of King Afonso's army.

The King's arrival with his fleets at Sagres, to join up with his uncle Prince Henry, was a memorable occasion. On the 3rd of October 1458 the royal ships rounded the Cape and joined those of Prince Henry, lying at anchor in Sagres Bay and further along the coast in the better harbour of Lagos. The combined fleets numbered some 220 ships; and the combined armies amounted to some 25,000 men. Because the expedition was so much nearer home than Constantinople, the King had been able greatly to increase the size of the army he had offered to the Pope. His army comprised several times as many men as Prince Henry had had under his command at Tangier, but was still probably smaller than the force with which King João I had captured Ceuta. However, it was an army in which the old divisions in the country were eradicated: Prince Pedro's son had been recalled from exile and was to fight beside his father's old antagonists. (King Afonso had attended the state interment of Prince Pedro's body at Batalha Abbey some years before, and had contributed to the restoration of his reputation.) There was also present the son of the old Duke of Braganza, who – as Count of Barcelos – had fought at Ceuta but now being over eighty had prudently remained at home.

King Afonso had never previously visited Sagres. Now he saw his uncle's strange Court, with its chart-rooms and instruments, its views across to Cape St Vincent and out to sea, its fortress walls and sheer cliffs, and – above all – its atmosphere of dedication to overseas exploration and adventure. The King's younger brother – another Prince Fernando – had been adopted as a son by Prince Henry on the eve of the Tangier expedition in 1436; in the subsequent twenty years he had, almost certainly, been a regular visitor to Sagres and was to grow ever closer to his ageing and childless uncle.

But although aged sixty-four, Prince Henry was far from a spent force. It was he who took effective command of the expedition. They crossed from Lagos to Tangier between the 14th and 19th of October and resisted the young King's impetuous suggestion that the plan should be changed in favour of attacking that desirable city. Two days later they landed on the beach in front of Alcácer Ceguer and – Prince Henry in the vanguard – drove the defending Moors back within their town walls.

Prince Henry had ensured that the siege equipment was more adequate for the task than had been the case at Tangier. A heavy cannon was landed and personally positioned by Prince Henry. He supervised a bombardment of the town walls throughout the night. Selecting one point on the walls, he consistently hammered it with the solid balls from the cannon, till eventually the wall cracked, crumbled and collapsed. With a breach clear into the town, the task of defence became hopeless.

The Moors sued for terms. Recalling the rigorous terms offered to his own army at Tangier in 1437, Prince Henry (in whose hands the King left the negotiations) must have been tempted to make punitive terms himself. But he seems to have resisted any temptation to vindictiveness. He offered the Moorish garrison a safe withdrawal from the town with their families and their portable belongings; he merely insisted that all Christian prisoners should be left behind. The Moors were told that the Portuguese wanted neither their lives nor their property: they wanted only that Alcácer Ceguer should become a part of Christendom.

When the Moors prevaricated and asked for time to consider the terms (and perhaps to repair the wall?) Prince Henry was firm: they must accept his terms immediately or he would put the town and all its inhabitants to the sword. This last was no more than normal practice where a town was taken by assault. The defenders did not hesitate further, but gratefully accepted the terms and began their evacuation. Prince Henry must have recalled the broken terms of the safe-conduct granted to him at Tangier; but he did not allow his own troops to molest the departing Moors. Throughout he had acted by those high standards of chivalry which had been instilled into him in early youth, and from which he had never deviated.

The entry of the Portuguese army into Alcácer Ceguer was triumphant. The fighting had been even less protracted and certainly less exacting than at Ceuta, but King Afonso outdid his grandfather in the distribution of honours and lands to his companions in arms. (His successor on the throne was to comment that the only lands he had inherited were the highways.) As after the fall of Ceuta, the principal mosque was quickly converted and consecrated and, even if the Pope could derive but small comfort from the service of mass at Alcácer while

Santa Sophia was echoing to the praises of Mohammed, Prince Henry must have felt that the disgrace of Tangier was expunged.

He urged the King to leave a strong garrison in the place, but his advice – now the fighting was over – was not heeded. King Afonso, Prince Henry and the main army sailed home to Portugal. No sooner had they left Morocco than the King of Fez promptly laid seige to the town. The Portuguese governor endeavoured to send a message to the rear-guard party of the Portuguese army to warn them that his supplies were too low to endure a long siege. A letter was shot by arrow and intended to fall beyond the besiegers; but it fell short and the King of Fez, learning from it of the plight of the garrison, offered favourable terms for an early surrender of the newly-won town. However, the governor was imbued with that spirit of gallantry which Prince Henry so frequently imparted to his associates; he replied by offering the King of Fez his own scaling ladders if he wished to make a fight of it. The King of Fez withdrew to gather further forces and returned with 30,000 cavalry and substantial quantities of infantry and artillery. Fifty-three days of further siege ensued at the conclusion of which the King of Fez abandoned the project. As he marched his troops away from the tiny and hungry garrison, the governor sent a message after him admonishing him not to give up so easily. Prince Henry, withdrawn once more at Sagres, was cheered by reports of this knightly conduct.

For now the last crusade, the last campaign was over. When he returned home in November 1458, Prince Henry himself seemed aware that his eventful and – for the fifteenth century – lengthy life was drawing to a close. His interest in the voyages down the African coast did not flag. He made good the promise of Diogo Gomes to send a priest to instruct Chief Nomimansa of the Gambia in the teachings of Christianity. He helped a promising navigator – Pedro de Cintra – to equip his caravels for a voyage which was to reach the coast of Sierra Leone too late for Prince Henry to learn of its success. He bequeathed the ecclesiastical

The fauna of the Guinea coast and the human inhabitants as depicted by early European travellers

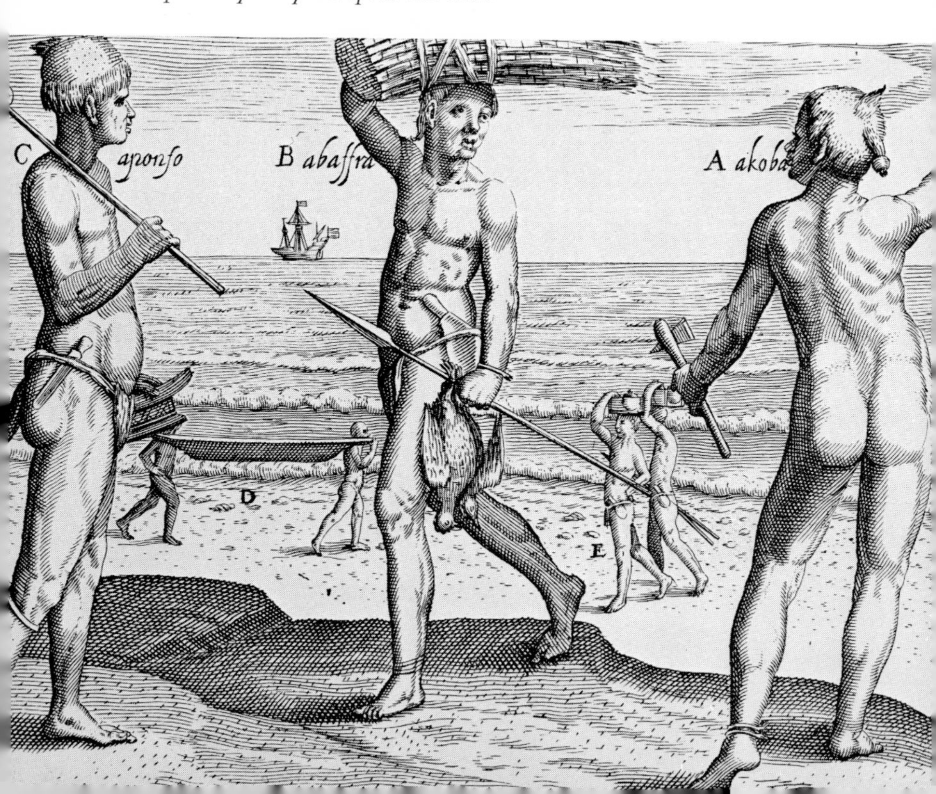

The Founder's Chapel of Batalha Abbey in which Prince Henry is buried together with his parents and brothers

revenues of Madeira and Porto Santo to the Order of Christ and the secular revenues to the Crown.

He also continued to take an active interest in the compilation of a world map in Venice, largely based on the results of the discoveries made by his sea captains. This remarkable work was undertaken at the Camaldolese Convent at Murano and enjoyed the patronage of the Doge. The map postulated a cape and triangular island at the southern extremity of Africa – 'Cavo di Diab' – which appears to have been based on reports of an Indian junk having rounded the cape from the east at the beginning of the century, searching for islands 'inhabited separately by men and women'. The Cape of Good Hope was not to be rounded by the Portuguese until twenty-six years after Prince Henry's death.

And now the end was near. One of the Prince's closest companions during his last months was Diogo Gomes, the sea captain who had been so successful in making peace with the chiefs of the Gambia and who is frequently referred to as his 'faithful servant'. It is fortunate that Diogo Gomes has left us his own account of Prince Henry's death, and it seems worth quoting* in full.

> In the year of our Lord 1460, Prince Henry fell ill in his town on Cape St Vincent, and of that sickness he died on Thursday, the 13th of November of the same year. And the same night which he died, they carried him to the church of St Mary in Lagos, where he was buried with all honour. At that time King Afonso was in Evora, and he, together with all his people, mourned greatly over the death of so great a Prince, when they considered all the expeditions which he had set on foot, and all the results which he had obtained from the land of Guinea, as well as how much he had laid out in continuous warlike armaments at sea against the Saracens in the cause of the Christian faith.

*From the English translation by Richard Henry Major in *The Life of Prince Henry the Navigator*, published in London in 1868 and still in print with Frank Cass & Co. Ltd.

At the close of the year King Afonso ordered me to be sent for, for, by the King's command, I had remained constantly in Lagos near the body of the Prince, giving out whatever was necessary to the priests, who were occupied in constant vigils and in Divine service in the church. And the King ordered that I should look and examine if the body of the Prince was decomposed, for it was his wish to remove his remains to the most beautiful monastery called Santa Maria de Batalha, which his father, King João I, had built for the Order of Friars Preachers. When I approached the body of the deceased, I found it dry and sound, except the tip of the nose, and I found him clothed in a rough shirt of horse-hair. Well doth the Church sing 'Thou shalt not give thine holy one to see corruption.' That my Lord the Infant (Infante) had remained a virgin till his death, and what and how many good things he had done in his life, it would be a long story for me to relate.

The King then issued a command that his brother Dom Fernando, Duke of Beja, and the bishops and nobles should go and convey the body to the aforesaid monastery of Batalha, where the King would await its arrival.

And the Prince's body was placed in a large and most beautiful chapel which King João his father had built, and where lie the bodies of the King and his Queen Philippa, the Prince's mother, together with his five brothers, the memory of all of whom is worthy of praise for evermore. There may they rest in holy peace. Amen.

Prince Henry's debts on his death were enormous: even twelve years earlier the Duke of Braganza had found it necessary to declare that Prince Henry owed him nearly twenty thousand gold crowns, and later he recorded that this sum had almost doubled. But to the last Prince Henry had been unconcerned by finances for their own sake: one of his last acts was to confirm his endowments at the University of Lisbon and to instruct that every Christmas twelve silver marks should be paid to the lecturer in theology. It seems that his interest in the commercial aspects of the later voyages was less related to any hopes of personal gain than to an understanding of the fact that trade would be the best and most enduring incentive to subsequent voyagers. It is pleasant to think that his debts were largely settled by his nephew and adopted son – Prince Fernando.

How should posterity judge the man who had died? After death he was to become many things to many men. He was to become a symbol of humanism to an age – a century after his own – that admired humanism. He was to become a symbol of scientific endeavour to those nineteenth-century writers who were absorbed by the advance of science. He was to become a shrewd Christian strategist, intent on turning the flank of Islam, to at least one recent Portuguese historian.* He was to become a symbol of colonial achievement to those Portuguese in our own age who have admired such achievement.

All these theories about him presuppose that he held to a single conscious objective or attitude. They tend to suggest that because Prince Henry's empirical methods opened the way to a more questioning frame of mind, this was his intention; that because his encouragement of the nautical sciences led to advances in this field, that was his motivation for such encouragement; because the campaigns in Morocco and – after his lifetime – the Portuguese presence in the Indian Ocean embarrassed or undermined the domination of Islam, that was the aim of his activities; because the new coasts he discovered and the trading station he set up were the beginnings of European colonial expansion, that was the vision towards which he strove.

Such theories surely overlook both the progressive development of Prince Henry's character and the conflicting nature of the influences affecting him. We have seen how, as a young man, he grew up in an excessively and self-consciously chivalric court. We have seen how these *mores* were crystallized in his consciousness by the exhilarating success – both personal and national – of the Ceuta expedition. We have observed the lingering effects of this leading to a bitter experience in the Tangier

*Joaquim Bensaude, *A cruzada do Infante D. Henrique*, Lisbon, 1943.

campaign and a distracting one in the Alcácer Ceguer venture. We have seen how, at various junctures in his life – notably during the fall of Prince Pedro – these old-fashioned chivalric notions led him into miscalculations and distorted his judgement.

But from an early period in his life we have also seen a set of quite different influences at work, conditioning his character and behaviour. We have noted that his sojourn in Africa, after the capture of Ceuta, opened his eyes to wider horizons – first across the Sahara desert and later across the untravelled waters of the Atlantic. As intelligence of freshly discovered islands, new opportunities for settlement, strange animals and plants, unfamiliar species of humanity, untapped sources of gold . . . and a score of other intriguing and stimulating phenomena were revealed to him, he systematically set about furthering the process of discovery. He lent his princely patronage to those who could aid the process, be they shipbuilders or cartographers, merchants or astronomers, seamen or courtiers. It was his own enthusiasm and determination which had pushed his sea captains out to the islands and around Cape Bojador; it was the intelligence which they brought back from their ventures which later fired Prince Henry to greater efforts still.

As these twin influences – his courtly upbringing and his zest for discovery – interacted, so he adapted and rationalized to bring into harmony with them the other main consideration in his life: his Christian faith. He quoted his religion repeatedly as the justification for all his undertakings, be they military expeditions in Morocco, slave-raiding down the African coast or venturing further into the unknown. And he was no hypocrite. The ascetic prince who wore a hair-shirt, eschewed wine and women and chose to live on a barren, windswept promontory did not consciously deceive himself. But he did possess that medieval capacity to bring his religion into everything he did and bring everything into his religion, to a degree at which motive and achievement, cause and effect were inexorably intertwined and, ultimately, fused. He explained his endeavours in theological terms and justified the results – even when they were unexpected – in the same terms.

Prince Henry's life had few of the gentler aspects that enrich most men's existence: he lacked close human relationships and domesticity, and one suspects he may not have been over-endowed with tenderness or humour. He was the very antithesis of a hedonist and yet he exemplified Pater's requirement of 'burning with a hard, gemlike flame'. The flame of his intense personality, for all its internal dichotomy and its muddled, mystic motivation, was yet to prove fierce enough to kindle the furnace of European exploration which was to burn for four centuries to come.

Epilogue

'Henceforth, wherever thou may'st roam,
My blessing, like a line of light,
Is on the waters day and night,
And like a beacon guards thee home.'

Lord Tennyson

The death of Prince Henry removed the immediate motivating force from the voyages of exploration down the African coast. For the next twenty years, until his own death in 1481, King Afonso V concentrated most of his ambitions and efforts on expeditions against the King of Fez in Morocco, thus earning for himself the title of 'the African'.

Encouraged by the success of the Alcácer Ceguer adventure, the King found it impossible to resist a further attempt on Tangier. In 1464, having heard tales from Portuguese prisoners of a disused sewer tunnel giving access to the city from outside the walls, he launched a series of attacks on it; the sewer route proved nonexistent and the attacks were repulsed. Some years later, however, King Afonso managed to capture the lesser town of Arzila and the inhabitants of Tangier, horrified by the ensuing slaughter and depressed by divisions within the ranks of Islam, surrendered their city to the Portuguese. The long-sought prize had been achieved; the bones of 'the martyred Prince', Fernando, were returned to his native land for burial at Batalha with his brothers.

While busy with these preoccupations, King Afonso could not mount and supervise voyages further south. In 1469, however, he farmed out the trade with the Guinea coast to a Lisbon merchant – Fernando Gomes. In return for trading concessions, Gomes was obliged by the terms of his contract to venture 300 miles further along the coast every year; in the course of the five years of his monopoly his agents had covered all the coastline from Sierra Leone to the Bight of Biafra. How disappointed Prince Henry would have been to learn that from there onwards the coast again turned due south: the Indian Ocean was farther off than he had known.

Meanwhile, King Afonso had quarrelled with Castile and led an abortive invasion campaign. He attempted to counteract his military failure with a diplomatic success, going to France to invite Louis XI to enter the war on his side. When this too failed, Afonso abdicated in disgust and took up the life of an itinerant friar: he disappeared into the French countryside. King Louis did not intend to add to his own problems that of a mendicant monarch at large in his realm; he had Afonso traced and returned to Portugal, where he duly reassumed the throne. Largely through the good offices of his son, peace with Castile was re-established.

During the war, the Castilians had begun raiding the Guinea coast and attacking Portuguese shipping. The peace treaties reiterated that the King of Portugal was Lord of the Guinea coast and of all the markets and mines thereon 'discovered or to be discovered', as well as Lord of Madeira, Porto Santo, the Azores and the Cape Verde Islands; on the other hand the Crown of Castile's claim to the Canaries was confirmed. It was agreed that Portugal would have exclusive rights to fight the Moors in the Kingdom of Fez (Morocco) and that Castile should have such rights in the Kingdom of Granada (on the Spanish mainland).

King Afonso V died in 1481 and was succeeded by his son King João II, who took a much more active interest in the West African coast. It was he who saw the need for another fortress trading station (on the same lines as that at Arguim) further down the coast, and he sent a largely prefabricated fort for erection at São Jorge da Mina, close to the site of the present city of Accra, on what was to be known for nearly five centuries as the Gold Coast. The captain to whom this enterprise was entrusted was required to use considerable diplomacy as the chosen site for the fort appeared to be a local graveyard; but good relations with the ruling chiefs were established and from then onwards gold began to flow steadily through this armed trading station to Portugal.

It was also during the early years of King João's reign that one of the most celebrated of his captains – Diogo Cão – discovered the Congo (Zaire) River. He erected the customary pillar (made of stone and carved with the royal arms of Portugal) at the mouth of the mighty river, and then sailed up its lower reaches to contact Manicongo, the most powerful African chief whom the Portuguese had so far encountered. He sent part of his crew through the forests to take greetings and gifts to Manicongo; when they failed to return by the expected date, he waited a while and then sailed home without them. It was three years before he again reappeared at the Congo River; his shipmates had survived and he pressed on down the coast with them past Angola to the latitude of Namibia (South-West Africa). Disappointed that the coast stretched on so far with no sign of the end of the continent, he turned back and made personal contact with Manicongo on the return voyage. The chief was friendly and declared his wish to be converted to Christianity. Diogo Cão had done sufficient to ensure for himself a place in the annals of Portuguese discovery.

Within a few months of Cão's return, King João sent out another of his captains – Bartolomeu Dias. It was realized that he would now be sailing so far from the fortress and supply base of Mina, and along such as inhospitable coast, that an expendable supply ship, to carry additional food and fresh water, was a necessary adjunct to his own caravel. As Dias sailed into waters further south than any so far reached, he encountered persistent head winds. He boldly sailed out into the southern Atlantic hoping to find westerlies to blow him to a point further south; in fact they blew him clean round the Cape of Good Hope. He sailed on just far enough to be sure that the coast turned northeastwards and that it was indeed the southernmost point of Africa which he had rounded, and then – after vain attempts to scrape up an acquaintance with some Hottentots – he too turned for home. The way now really was clear for the opening up of the sea route to the Indies. But it was not immediately pursued.

King João had another overriding preoccupation. One of Prince Henry's original motives for launching the first voyages down the African coast had been to make contact with the kingdom of that mysterious and legendary Christian monarch – Prester John. For centuries this enigmatic figure had haunted European imaginations. In the twelfth century a letter, purporting to be addressed by Prester John to the Emperor Manuel, had circulated widely in Europe. Copies of the letter, in various forms, proliferated and later historians traced as many as a hundred (of which eight are in the British Museum). The letter claimed that its author was 'by the power and virtue of God and the Lord Jesus Christ, Lord of Lords and the Greatest Monarch under heaven'. His realm was somewhat vaguely situated in furthest Asia, and included in its territory the Tower of Babel; seventy-two kings, ruling over as many kingdoms, were his tributaries. Extraordinary beasts inhabited his domains: monstrous ants that dug up gold, fish that emitted imperial purple dye, salamanders that lived in fire. The rivers abounded with rare gems, including magical pebbles that restored sight to the blind. His palace had a vast mirror erected in front of it in which Prester John could see all that transpired within his dominions and could detect conspiracies in his furthest provinces. The palace itself was made of ebony and crystal; a roof of precious stones was supported on columns of purest gold. All these extraordinary claims appear to have been taken, if not altogether literally, at least as having as much validity (which they did) as the later and accepted travelogues of Sir John Mandeville.

By the fifteenth century, however, it had been generally agreed that the kingdom of Prester John was not in 'furthest Asia' but in Ethiopia – an almost equally unknown region and also loosely described as 'the Indies' by contemporary geographers. Making contact with Prester John had been a subsidiary objective for most of Prince Henry's expeditions – an additional dividend which might at any time accrue from the voyages of discovery.

In fact, when each great river was discovered on the West African coast, there was renewed speculation about whether it might not be that 'Western Nile' which would lead to Ethiopia and the Kingdom of Prester John. This was particularly the case with the discovery of the Senegal River.

But it was not until the reign of King João II that any persistent effort was made to establish contact. The prospect of a strategically-placed Christian ally – even if he was a Coptic schismatic – was ever more alluring as the chances of reaching the Indies increased. In 1483, one of the Portuguese sea captains reported that the Kings of Benin (ruling a part of what is now Nigeria) used to send envoys and gifts to an overlord whose capital was twleve months' journey further east. Such a journey might well reach to Ethiopia. Among the insignia sent back by this overlord were small crosses. Could this be Prester John?

King João decided to attempt an approach from the Egyptian and Arabian direction. Overland travel through these Moslem lands was not unknown and those Ethiops who had appeared in Europe had all come by that route. In fact, a number of monks and envoys had reached Europe in the fifteenth century; Rome, Venice and even Aragon had been visited by Christian emissaries from Ethiopia. King João sent two intrepid travellers to Alexandria, where they disguised themselves as Arab merchants. One of them – Pêro de Covilhã – sailed through the Red Sea to India and spent more than a year studying the markets in spices, gold and precious stones. He returned to Cairo, having also seen something of the east coast of Africa, only to find that King João had sent letters to him ordering him not to return home until he had completed his instructions by 'reaching the Kingdom of Prester John'. Pêro de Covilhã wrote a full report for the King of his journeyings so far (a report which must have proved invaluable to Vasco da Gama a few years later) and then loyally set out in 1490 southwards to Ethiopia. Nothing is known of the route he took but he is recorded as having indeed found the court of the Ethiopian Emperor Iskander, on the shores of Lake Tana. He was well received, but when the Emperor died his successor decided that contacts with the King of Portugal were of dubious value, and Pêro de Covilhã was provided with a wife and an estate but forbidden to leave Ethiopia. There he survived over thirty years before he died. Contact had been made, but the enthusiastic alliance with a Christian monarch whose domain bordered the Indian Ocean had hardly been forthcoming.

While this curious mission was under way, an event of far greater importance was taking place elsewhere. In 1484, Christopher Columbus had sought the patronage of King João II for a voyage of exploration across the Atlantic which, he confidently hoped, would prove the direct and quickest way to the Indies. He had almost certainly already visited the Portuguese settlements at Mina and on Madeira and the Azores. (It will be remembered that his father-in-law had been an original settler of Porto Santo.) Columbus was adamant in requiring three caravels and the title of Viceroy in all the lands he might discover. King João and his advisers were sceptical about the distance across the Atlantic Ocean; though they did not doubt that the route would eventually reach China and the Indies they (rightly) concluded the distance was far greater than Columbus calculated. Also their investment in the route round Africa was very heavy and appeared (again rightly) to be about to be crowned with success. Columbus was sent away and succeeded in offering his services to Ferdinand and Isabella in Castile. On the 4th of March 1493, King João and the citizens of Lisbon had the humiliating experience of seeing him sail up the Tagus having completed his mission and having discovered the New World under Spanish auspices.

These discoveries added urgency to the question of drawing up a demarcation line between Spain and Portugal in the New World (even if the latter were still thought to be the old Indies). The Pope decreed a line from north to south one hundred leagues west of the Azores: all land discovered or to be discovered west of that line was to be Spanish, and east of it Portuguese. King João did not accept this ruling but managed to negotiate with the Spanish Crown a line 370 leagues west of the Cape Verde Islands to which the same significance would be attached. This was embodied in 1494 in the Treaty of Tordesillas.

Brazil was not officially discovered until Pedro Álvares Cabral landed there in 1500. But there is considerable circumstantial evidence to suggest that the Portuguese had at least sighted land in the southern Atlantic well before that date. Indeed, the fact that the Treaty of Tordesillas brought western Brazil within the Portuguese domain may not have been coincidental. The boldly imaginative south Atlantic use of trade winds and currents made by Vasco da Gama a few years later also suggests some fairly extensive prior knowledge on the part of the Portuguese.

It was not only in the southern Atlantic that Portuguese captains had been sailing into uncharted oceans. There is reason to believe that the Portuguese fishing fleet first started bringing back catches from off the Newfoundland banks at about this time. Certainly they were well established in the area by the following century. When King João II died in 1495, although he had only reached the age of forty, he had prepared the way for his successor to grasp the prizes which he and his predecessors had sought so long.

King Manuel I considered that he had inherited from his great-uncle Prince Henry the sacred mission of exploration. He selected Vasco da Gama to command an expedition of four ships; they were no longer caravels, but square-rigged on the fore and main masts and only lateen-rigged on the mizzen mast, because the adverse return winds along the north-west African coast were no longer the main hazard which they expected. The ships were equipped with three years' supply of provisions, all the most modern instruments and considerable armament. Only in the matter of gifts for the natives did the King's imagination let him down: he supplied cheap trinkets – beads, bells and brass baubles – which were perfectly acceptable to African tribesmen but quite inappropriate for sophisticated Indian princes and merchants. Vasco da Gama received from the King a banner bearing the cross of the Order of Christ; the same emblem, under which Prince Henry's captains had also ventured forth, was on his sails. He spent his last night ashore in the tiny Restelo chapel, above Belem, which Prince Henry had built for his own mariners.

The story of Vasco da Gama's famous voyage has been retold many times, most memorably by the Portuguese epic poet Camões. From the Guinea coast, Vasco da Gama sailed out westward into the Atlantic, only turning eastward again to take advantage of the prevailing winds off the Brazilian coast. These carried him directly to the Cape of Good Hope. Once into the Indian Ocean he pressed on up the coast, giving the name of Natal to that part of the South African coast which they were passing at Christmas 1497. When he reached Mozambique Island, which was to become a Portuguese stronghold for many centuries, he was aware of entering the zone of Arab influence. He had troubles and fights there with the inhabitants and, further north at Mombasa, he would have fallen into an ambush had not one of the Arabs on board revealed the plot – after a judicious application of boiling oil – to Vasco da Gama. Only at Malindi, a little further north, was Vasco da Gama able to establish good enough relations with the Sultan for the latter to provide him with a pilot for the final stage of his voyage – across the Indian Ocean to Calicut on the west coast of India itself. He made his landfall on the 18th of May 1498.

Having reached their promised land of India, the Portuguese were determined to maximize the profits of the spice and other trades accruing from that region. To do this it was necessary to dominate the trade by sea power based on a few secure settlements. The philosophy of the fortress trading station – which bore fruit at Arguim and Mina – was now to be extended and developed. The Portuguese were fortunate in that they arrived in the Indian Ocean at a time when most of the more powerful rulers of Egypt, Arabia, Persia and India happened either to be in disarray or at loggerheads with each other. The Portuguese were not slow to exploit these factors to their own advantage. Nor were they hesitant about facing violent confrontations. As with their West African experience, a period of ruthlessness and fighting appeared a necessary prelude to a longer period of profitable trading.

In 1509, Francisco de Almeida destroyed the Egyptian-Gujarati fleet of Diu, thus eliminating the only Moslem naval force in the Indian Ocean which was a real threat to Portuguese domination. Afonso de Albuquerque then set about capturing the bases which the Portuguese needed to sustain control of the sea routes. In 1510 he seized Goa, as a foothold on the western coast of India, from the Sultan of Bijapur. In 1511, he captured Malacca, thus acquiring not only a major distribution centre for Indonesian spices, but control of the vital straits between the Indian Ocean and the seas of Java and South China. In 1515, he seized Ormuz Island at the entrance to the Persian Gulf; this was already an important clearing house for trade – particularly in spices, Arabian horses, gold and other goods passing between Persia and India – and when fortified by the Portuguese was a key-stone in their policy of controlling the arteries of commerce. Albuquerque's only failure was to control the entrance to the Red Sea; he attempted to capture Aden but without success, and some eastern trade continued to filter back to Europe via the Moslem-dominated route to Suez and Alexandria.

From the Indian Ocean, the Portuguese pressed on into the China seas. Here again they were fortunate that the Ming dynasty rulers of China had placed a prohibition on trade with the Japanese, whom they described as 'dwarf-robbers', in their own or Japanese ships. It was not long before the Portuguese had managed to acquire a virtual monopoly of trade between these two countries, to add to the trade which they were doing on their own behalf. They did not, however, manage to establish that degree of overall domination of commerce in the China seas which they had achieved in the Indian Ocean. Nevertheless, a permanent trading station was found necessary and in 1557 they established a foothold in Macao. It was twenty years later when the Emperor of China, residing at Peking, discovered how firmly installed they had become; by then there was nothing he could do but sanction the arrangement, and there the Portuguese have remained to this day.

How had so much been achieved in the hundred years since Prince Henry's death? How had Portuguese naval strength been maintained in the distant waters of the Far East and India when there was only one really adequate Portuguese dockyard (at Goa) in that whole quarter of the globe? How had Portuguese entrepreneurs successfully contested power with the sophisticated armies of the Orient – to whom firearms and cannon were no strangers by the sixteenth century – and simultaneously set about the colonization of Brazil and the exploration of the North Atlantic fisheries? How had a tiny, until recently poor, underpopulated country on the fringes of Europe managed to open up two thirds of the world to European commerce and to dominate that commerce herself?

No one answer to these questions can suffice. But if one clue only can be given to the answer, that clue surely lies in the character of the man who – more than any other – set this process in train. It was the Portuguese will to succeed which was their greatest single asset, that spirit of determination which – Sir George Sansom has pointed out – was stronger than the will of the Asian peoples to resist. And that spirit was the legacy of the hard, tough, not very lovable man who – coupling in his own personality the dedication of the crusader with the tenacious curiosity of the modern – has become known to the world by the inappropriate name of Prince Henry the Navigator.

Bibliography

The two volumes of the *Bibliografia Henriquina*, published in Lisbon in 1960 to commemorate the 500th anniversary of Prince Henry's death, extend to over 700 pages and weigh nearly seven pounds. From this it will be clear that every biographer of Prince Henry must be highly selective in his reading. I have consulted the main chronicles of the period and those recent Portuguese, English and French works which have seemed to throw new light on subjects related to Prince Henry and his times. I have also read a number of established earlier biographies and if echoes of their authors' thoughts or phrases have crept into my writing, I can only apologize for this unintentional tribute to the persuasiveness of their views or the aptness of their expression.

I considered listing after each chapter the works which I had used in its preparation, but the overlap would have been so great that it seemed best to provide one consolidated list of my main sources. In the case of the chronicles and other early writings, the dates of first publication are frequently in doubt and I have given instead the dates of the authors; in the case of more recent works I have attempted to provide the place and date of publication.

ÁLVARES, JOÃO (1410–1472 approx.), *Crónica do Infante Santo D. Fernando*

AZURARA *see* Zurara

BARROS, JOÃO DE (1496–1570), *Decada I*

BEAZLEY, C. RAYMOND, *Prince Henry the Navigator* (London 1895)

BELL, CHRISTOPHER, *Portugal and the Quest for the Indies* (London 1974)

BOVILL, E. W., *The Golden Trade of the Moors* (London 1958)

BRAGANZA, JAIME DUKE OF, ed. *Alguns Documentos do Arquivo Nacional da Torre do Tombo* (Lisbon 1892)

BOXER, C. R., *The Portuguese Seaborne Empire 1415–1825* (London 1969)

BRADFORD, ERNLE, *Southward the Caravels* (London 1961)

BRÁSIO, PADRE ANTÓNIO, *A Acção Missionária no Período Henriquino* (Lisbon 1958)

BROCHADO, COSTA, *Descobrimento do Atlântico* (Lisbon 1958)

CA DA MOSTO *see* Cadamosto

CADAMOSTO, LUÍS DE (1432–1477), *Voyages*, Hakluyt Society (London 1938)

CAMÕES, LUÍS VAZ DE (1524–1580), *Os Lusíadas*

CORTESÃO, A., *Cartografia e Cartógrafos Portuguesas* (Lisbon 1935)

DUARTE, D. (1391–1438), *O Leal Conselheiro*

FONSECA, QUIRINO DA, *Os Navios do Infante D. Henrique* (Lisbon 1958)

FONTOURA DA COSTA, A., *A Ciência Náutica dos Portugueses na Época dos Descobrimentos* (Lisbon 1958)

FOSS, MICHAEL, *Chivalry* (London 1975)

GODINHO, V. MAGALHÃES, *Documentos Sobre a Expansão Portuguesa Vol. I, II & III* (Lisbon 1945)
A Economia dos Descobrimentos Henriquinos (Lisbon 1962)
Os Descobrimentos e a Economia Medieval, 2 vols (Lisbon 1965)

GOMES, DIOGO (1420–1480 approx.), *Relações do Descobrimento da Guiné e das Ilhas dos Açores, Madeira e Cabo Verde*

GUINGUAND, MAURICE, *L'Or des Templiers* (Paris 1973)

HITCHINS, H. L. and MAY, W. E., *From Lodestone to Gyro-Compass* (London 1952)

HUIZINGA, J., *The Waning of the Middle Ages* (London 1924)

LIVERMORE, H. V., *A History of Portugal* (Cambridge 1947)

LOPES, FERNÃO (1370–1460), *Crónica de D. João I*

MAJOR, RICHARD HENRY, *The Life of Prince Henry the Navigator* (London 1868)

MARQUES, A. H. DE OLIVEIRA, *Guia do Estudante de História Medieval Portuguesa* (Lisbon 1964)

MARTINS, J. P. OLIVEIRA, *Os Filhos de D. João I* (Lisbon 1901)

MORISON, ADMIRAL SAMUEL ELIOT, *The European Discovery of America – The Southern Voyages 1492–1616* (New York 1974)

PALACIO, DIEGO GARCIA DE, *Instrucción Náutica para Navegar* Original ed. Mexico 1587 (Madrid 1944)

PARRY, J. H., *The Discovery of the Sea* (London 1974)

PINA, RUY DE (1440–1521), *Crónica de D. Duarte*
Crónica de D. Afonso V

PRESTAGE, EDGAR, *The Portuguese Pioneers* (London 1939)

RENAULT, GILBERT, *Les Caravelles du Christ* (Paris 1956)

RUNCIMAN, SIR STEVEN, *History of the Crusades*, vols II & III (London 1952)

RUSSELL, P. E., *English Intervention in Spain and Portugal in the Time of Edward III and Richard II* (Oxford 1955)
Prince Henry the Navigator, Canning House Lecture (London 1960)

SÁ, A. MOREIRA DE, *O Infante D. Henrique e a Universidade* (Lisbon 1960)

SANCEAU, ELAINE, *Henry the Navigator* (London 1946)

TAYLOR, E. G. R., *The Haven-Finding Art* (London 1956)

TEIXEIRA DA MOTA, CAPTAIN A., *Evolução dos Roteiros Portugueses* in *Publicações do Agrupamento de Estudos de Cartografia Antiga XXXIII* (Coimbra 1969)

ZURARA, GOMES EANES DE (1410–1473), *Crónica da Tomada de Ceuta*
Crónica do Descobrimento de Conquista da Guiné

Index